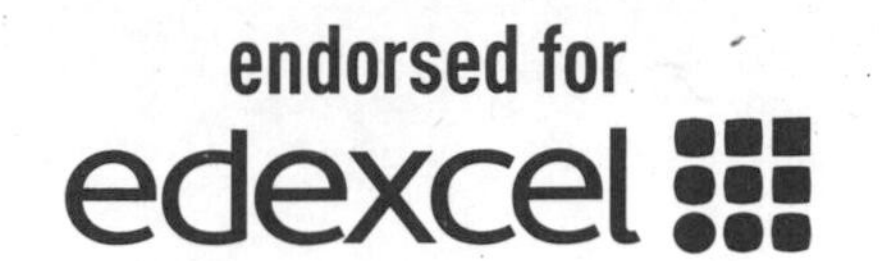

REVISE EDEXCEL GCSE

Mathematics

Specification A Linear

REVISION WORKBOOK

Higher

Series Director: Keith Pledger

Series Editor: Graham Cumming

Authors: Authors: Julie Bolter, Gwenllian Burns, Jean Linsky

A note from the publisher

In order to ensure that this resource offers high-quality support for the associated Edexcel qualification, it has been through a review process by the awarding body to confirm that it fully covers the teaching and learning content of the specification or part of a specification at which it is aimed, and demonstrates an appropriate balance between the development of subject skills, knowledge and understanding, in addition to preparation for assessment.

While the publishers have made every attempt to ensure that advice on the qualification and its assessment is accurate, the official specification and associated assessment guidance materials are the only authoritative source of information and should always be referred to for definitive guidance.

Edexcel examiners have not contributed to any sections in this resource relevant to examination papers for which they have responsibility.

No material from an endorsed resource will be used verbatim in any assessment set by Edexcel.

Endorsement of a resource does not mean that the resource is required to achieve this Edexcel qualification, nor does it mean that it is the only suitable material available to support the qualification, and any resource lists produced by the awarding body shall include this and other appropriate resources.

Contents

A small bit of small print

A grade allocated to a question represents the highest grade covered by that question. Sub-parts of the question may cover lower grade material.

The grade range of a topic represents the usual grade range that the topic is assessed at. The topic may form part of a higher grade question if tested within the context of another topic.

Questions in this book are targeted at the grades indicated.

Factors and primes

1 (a) Express the following numbers as products of their prime factors.

Guided

(i) 60

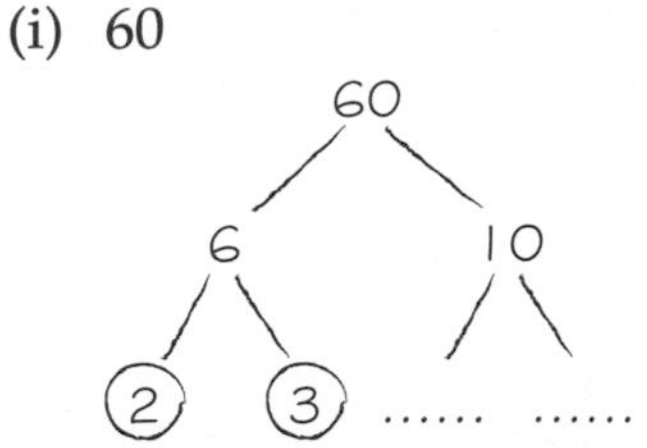

(ii) 150

150

10 15

2 ....

Remember to circle the prime factors as you go along.

$60 = 2 \times ...... \times ...... \times ......$ **(2 marks)**

$150 = 2 \times ...... \times ...... \times ......$ **(2 marks)**

(b) Find the highest common factor (HCF) of 60 and 150

Guided

$60 = 2 \times 3 \times ...... \times ......$

$150 = 2 \times ...... \times ...... \times ......$

$HCF = 2 \times ...... \times ......$

$= ....$

Circle all the prime numbers which are common to both products of prime factors. Multiply the circled numbers together to find the HCF.

(1 mark)

(c) Find the lowest common multiple (LCM) of 60 and 150

Guided

$LCM = ... \times ...... \times ......$

$= ....$

To find the LCM, multiply the HCF by the numbers in both products that were not circled in part (b).

(1 mark)

C 2 (a) Express 72 as a product of its prime factors.

.... **(2 marks)**

(b) Find the highest common factor (HCF) of 72 and 120

HCF = ... **(1 mark)**

(c) Find the lowest common multiple (LCM) of 72 and 120

LCM = ... **(1 mark)**

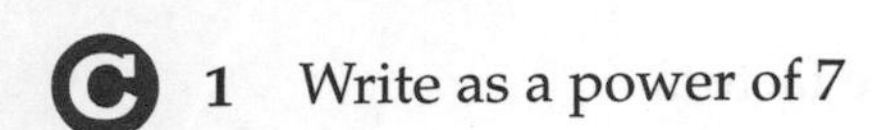

Indices 1

C 1 Write as a power of 7

(a) $7^3 \times 7^{10}$

Guided $7^3 \times 7^{10} = 7^{3+10} = \ldots 7^{13} \ldots$ **(1 mark)**

(b) $7^{15} \div 7^9$

Guided $7^{15} \div 7^9 = 7^{15-9} = \ldots 7^6 \ldots$ **(1 mark)**

(c) $\frac{7^{12}}{7^4 \times 7}$

Guided $\frac{7^{12}}{7^4 \times 7} = \frac{7^{12}}{7^4 \times 7^1} = \frac{7^{12}}{7^{8}}$

$= \ldots 7^4 \ldots$ **(2 marks)**

(d) $(7^5)^4$

Guided $(7^5)^4 = 7^{5 \times 4} = \ldots 7^{20} \ldots$ **(1 mark)**

C 2 Write as a power of 5

(a) $5^8 \times 5^4$ (b) $\frac{5^{12} \times 5}{5^4 \times 5^3}$ $\frac{5^{13}}{5^7}$ (c) $(5^2)^3$

$\ldots 5^{12} \ldots$ **(1 mark)** $\ldots 5^6 \ldots$ **(2 marks)** $\ldots 5^6 \ldots$ **(1 mark)**

C 3 $6^8 \times 6^3 = 6^5 \times 6^x$

Find the value of x.

Use the index laws to simplify each side of the equation.

$6^8 \times 6^3 = 6^{12}$

$6^5 \times 6^7 = 6^{12}$

$x = \ldots 7 \ldots$ **(2 marks)**

B 4 Simplify 4^0

$\ldots 1 \ldots$ **(1 mark)**

A 5 Write $9^3 \times 27^2$ as a single power of 3

Guided $9^3 \times 27^2 = (3^{\cdots\cdots})^3 \times (3^{\cdots\cdots})^2$

$= 3^{\cdots\cdots} \times 3^{\cdots\cdots}$

$= \ldots\ldots\ldots\ldots$ **(2 marks)**

A 6 Write $8^6 \div 4^3 \times 2^5$ as a single power of 2

$\ldots\ldots\ldots\ldots$ **(2 marks)**

Fractions

C **1** Work out $3\frac{2}{3} + 1\frac{4}{5}$

Give your answer as a mixed number in its simplest form.

Guided

$3\frac{2}{3} + 1\frac{4}{5}$

$= \dots\dots + \frac{\dots}{3} + \frac{\dots}{5}$ — Add the whole numbers.

$= \dots\dots + \frac{\dots}{15} + \frac{\dots}{15}$ — Write as equivalent fractions with the same denominator.

$= \dots\dots + \frac{\dots}{15}$

$= \dots\dots + 1\frac{\dots}{15}$

$= \dots\dots\frac{\dots}{15}$ — Write your final answer as a mixed number in its simplest form. **(3 marks)**

C **2** Work out (a) $7\frac{1}{7} - 2\frac{2}{3}$ (b) $8\frac{9}{10} + 2\frac{3}{5}$

Give each answer as a mixed number in its simplest form.

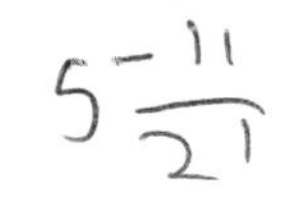

(a) .... **(3 marks)** (b) .... **(3 marks)**

C **3** Work out $2\frac{1}{3} \times 1\frac{3}{5}$

Exam questions similar to this have proved especially tricky – be prepared! ResultsPlus

Guided

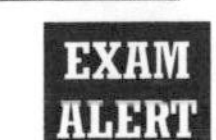

$2\frac{1}{3} \times 1\frac{3}{5}$

$= \frac{\dots}{3} \times \frac{\dots}{5}$ — Write both mixed numbers as improper fractions.

$= \frac{\dots}{\dots}$ — Multiply numerators **and** multiply denominators.

$= \dots\dots\frac{\dots}{\dots}$ — Write your final answer as a mixed number in its simplest form. **(3 marks)**

C **4** Work out (a) $2\frac{1}{4} \times 3\frac{1}{3}$ (b) $5\frac{1}{3} \div 1\frac{2}{9}$

Give each answer as a mixed number in its simplest form.

(a) .... **(3 marks)** (b) .... **(3 marks)**

C **5** Work out (a) $8\frac{5}{6} - 3\frac{2}{5}$ (b) $4\frac{1}{5} \div \frac{9}{10}$

Give each answer as a mixed number in its simplest form.

(a) .... **(3 marks)** (b) .... **(3 marks)**

Decimals

C **1** Using the information that $67 \times 29 = 1943$
write down the value of

(a) 6.7×2.9

Guided $6.7 \times 2.9 = 1943 \div$ …………
$=$ …………

67 has been divided by 10 and 29 has been divided by 10. So the answer needs to be divided by 100.

(1 mark)

(b) 670×0.0029

Guided $670 \times 0.0029 = 1943 \div$ …………
$=$ …………

67 has been multiplied by 10 and 29 has been divided by 10 000. So the answer needs to be divided by 1000.

(1 mark)

(c) $19\,430 \div 67$

Guided $19\,430 \div 67 = 29 \times$ …………
$=$ …………

1943 has been multiplied by 10 and 67 is unchanged. So multiply 29 by 10.

(1 mark)

C **2** Use the information that $127 \times 84 = 10\,668$
to find the value of

(a) 1270×84 (b) 0.127×8.4 (c) $10\,668 \div 1.27$

………………… **(1 mark)** ………………… **(1 mark)** ………………… **(1 mark)**

C **3** Given that $63 \times 48 = 3024$
write down the value of

(a) 6300×4.8 (b) 0.063×4.8 (c) $30\,240 \div 6.3$

………………… **(1 mark)** ………………… **(1 mark)** ………………… **(1 mark)**

Recurring decimals

1 Express $0.\dot{1}\dot{5}$ as a fraction in its simplest form. You must use algebra.

Guided

Let $x = 0.\dot{1}\dot{5}$

$100x = 15.151515...$

$-\quad x = 0.151515...$

$99x = 15$

$x = \frac{15}{99}$

$x = \frac{5}{33}$

(3 marks)

A **2** Change the recurring decimal $0.\dot{8}$ to a fraction. You must use algebra.

$x = 0.88888....$

$10x = 8.888....$

$9x = 8$

$x = \frac{8}{99}$

$\frac{8}{99}$ **(2 marks)**

A **3** Convert the recurring decimal $2.\dot{4}1\dot{7}$ to a fraction. You must use algebra.

$x = 2.417417417....$

$1000x = 2417.417417....$

$999x = 2415 \quad \frac{2415}{999} = \frac{805}{333}$

.... **(3 marks)**

A **4** Convert the recurring decimal $0.4\dot{7}$ to a fraction. You must use algebra.

Guided

Let $x = 0.4\dot{7}$

$10x = 4.7777777...$

$-\quad x = 0.4777777...$

$9x = 4.......$

$x = \frac{.........}{9}$

$x = \frac{.........}{.........}$

Multiply the top and bottom of the fraction by 10.

(3 marks)

A **5** Prove that $0.8\dot{2}\dot{7}$ can be written as the fraction $\frac{91}{110}$

(3 marks)

Rounding and estimation

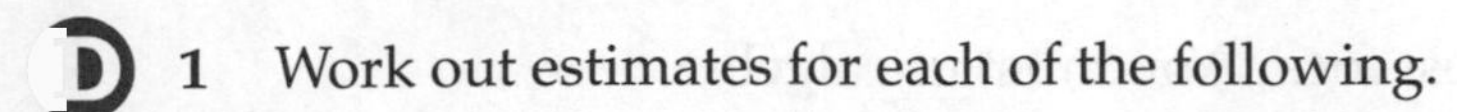

D **1** Work out estimates for each of the following.

(a) 145×78

Guided

$100 \times$ .... = ...

(1 mark)

(b) $19.1 \div 1.51$

Guided

.... $\div\ 2$ = ...

(1 mark)

(c) $48.9 \times 2.78 \times 11.9$

Guided

.... $\times$ .... $\times\ 10$ = ...

(1 mark)

D **2** Work out an estimate for the value of $\dfrac{3981}{2.3 \times 18.7}$

.... **(2 marks)**

C **3** Work out an estimate for the value of $\dfrac{612 \times 39}{0.53}$

.... **(2 marks)**

C **4** Work out an estimate for the value of $\dfrac{40.7 \times 1.6}{0.053}$

Exam questions similar to this have proved especially tricky – be prepared! ResultsPlus

Guided

EXAM ALERT

$\approx \dfrac{40 \times \text{.........}}{0.05} = \dfrac{\text{.........}}{\text{.........}}$

First round each number to 1 significant figure.

= ....

(2 marks)

C **5** Work out an estimate for the value of $\dfrac{9.73 \times 4.12}{0.0214}$

.... **(2 marks)**

C **6** Work out estimates for the following calculations.
State whether your answer is an underestimate or an overestimate.

(a) $\dfrac{995.3}{5.3 \times 11.3}$

.... **(2 marks)**

(b) $\dfrac{101.7}{3.7 \times 4.72}$

.... **(2 marks)**

C **7** Work out an estimate for the value of 2.52×11.7^2

.... **(2 marks)**

Upper and lower bounds

1 The length of a rectangle is 9.7 cm correct to 2 significant figures.
The width of the rectangle is 6.5 cm correct to 2 significant figures.
Work out the upper bound for the area of the rectangle.

Guided

Upper bound of length = 9.75

Upper bound of width = ...

Upper bound of area = 9.75 × ....

= ....

= .... cm^2

(3 marks)

2 The length of a rectangle is 24 cm correct to 2 significant figures.
The width of the rectangle is 9.6 cm correct to 2 significant figures.
Work out the lower bound for the perimeter of the rectangle.

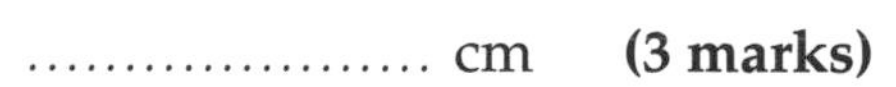

.... cm **(3 marks)**

3 A ball is dropped from a window.
The time that it takes to reach the ground is given by the formula $t = \sqrt{\frac{2s}{a}}$
where a m/s^2 is the acceleration due to gravity and s m is the height of the window.
$s = 117$ m correct to 3 significant figures
$a = 9.8$ m/s^2 correct to 2 significant figures

(a) Calculate the lower bound and the upper bound for the value of t.
Give your answers correct to 4 decimal places.

.... **(4 marks)**

(b) Use your answers to part (a) to write down the value of t to a suitable degree of accuracy.
You must explain your answer.

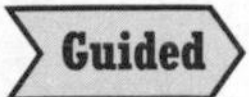

t = ... because the upper bound and the lower bound both agree to 2 significant figures

(1 mark)

Fractions and percentages

D **1** Uzma invests £4000 in a bank account for 1 year.
Interest is paid at a rate of 2.5% per annum.
How much interest will Uzma get at the end of 1 year?

Guided $\frac{\dots\dots}{100} \times £4000 = £\dots\dots\dots\dots$ **(2 marks)**

D **2** A farmer has 48 llamas.
30 of the llamas are female.

(a) Work out 30 out of 48 as a percentage.

Guided $\frac{30}{\dots\dots} \times 100 = \dots\dots\dots\dots$ **(2 marks)**

60% of the female llamas are pregnant.

(b) Write the number of pregnant female llamas as a fraction of the 48 llamas.
Give your answer in its simplest form.

.... **(2 marks)**

D **3** Meera works in an electrical shop.
Each week she gets paid £160 plus 15% of the value of the goods she sells.
One week Meera sold £3200 of goods.
Work out the total amount she was paid this week.

£.................... **(3 marks)**

D **4** Liam's annual income is £16 000
He pays $\frac{1}{5}$ of the £16 000 in rent.
He spends 15% of the £16 000 on food.
Work out how much of the £16 000 Liam has left.

£.................... **(4 marks)**

C **5** At an outdoor centre, 140 students each choose one activity.
$\frac{1}{7}$ of the students choose rock climbing.
$\frac{3}{7}$ of the students choose rafting.
All the rest of these students choose abseiling.
How many students choose abseiling?

.... **(3 marks)**

Percentage change

1 A washing machine costs £420 plus 20% VAT.
Calculate the total cost of the washing machine.

Guided

VAT = $\frac{20}{100}$ × ....

= ....

Total cost = 420 + ...

= £.....................

(3 marks)

D **2** Helen buys a jacket in a sale.
The normal price is £84
The normal price of the jacket is reduced by 35%.
Work out the sale price of the jacket.

£..................... **(3 marks)**

C **3** Eliza went to New York.
She changed pounds (£) into American dollars ($).
The exchange rate was £1 = $1.60
The value of the pound has decreased from $1.60 to $1.56
Calculate the percentage decrease in the value of the pound.

Guided

Percentage decrease = $\frac{\text{decrease}}{\text{original value}} \times 100$

= $\frac{.........}{.........} \times 100$

= ....%

(3 marks)

C **4** Ali buys 120 cans of drink for a total of £30
He wants to make a profit of 40%.
Work out the price for which he should sell each can of drink.

.... **(4 marks)**

C **5** Jean books a holiday.
The total cost of the holiday is £1430
She pays a deposit of 35% of the total cost.
She pays the rest in 10 monthly instalments.
Work out how much she pays each month.

£..................... **(4 marks)**

Reverse percentages and compound interest

B **1** Linda bought a new car for £18 000
Each year, the car depreciated in value by 15%.
Work out the value of the car after 4 years.

Guided Multiplier $= 1 - \frac{15}{100} =$ ....

Work out the multiplier as a decimal.

Value after 4 years = 18 000 × (.........)^4

= ....

= £.....................

When working with money, answers must be given to 2 decimal places.

(3 marks)

B **2** Jalin invested £3200 in a savings account for 3 years.
He was paid compound interest at a rate of 3.5% per annum.
Work out how much was in the account after 3 years.

£..................... **(3 marks)**

B **3** In a sale, normal prices are reduced by 35%.
The sale price of a DVD player is £403
Work out the normal price of the DVD player.

EXAM ALERT

Exam questions similar to this have proved especially tricky – be prepared! ResultsPlus

Guided Multiplier $= 1 - \frac{35}{100} =$ 0.65

Normal price = 403 ÷ ~~620~~ 0.65

= £ 620

(3 marks)

B **4** Jill's weekly pay this year is £460
This is 15% more than her weekly pay last year.
Dave says, 'This means Jill's weekly pay last year was £391.'
Dave is wrong. Explain why.

(2 marks)

A **5** Pete invested £5100 for n years in a savings account.
He was paid 4.5% per annum compound interest.
At the end of the n years he had £6641.53 in the savings account.
Work out the value of n.

Choose some values for n and work out the amount in the savings account after n years.

$n =$ .... **(2 marks)**

Ratio

D 1 There are 60 toy cars in a box.
18 of the toy cars are blue.
The rest of the toy cars are red.
Write down the ratio of the number of red toy cars to the number of blue toy cars.
Give your ratio in its simplest form.

Guided

Number of red cars = 60 − ....

= ....

Make sure you put the numbers in the ratio in the correct order.

Ratio of red cars to blue cars = ...:.........

= ....:......... **(2 marks)**

D 2 There are 32 students in a class.
20 of the students are girls.
Rosie says, 'The ratio of the number of girls to the number of boys in this class is 3:5.'
Is Rosie right?
You must give a reason for your answer.

.... **(2 marks)**

D 3 Ahmed and James share £120 in the ratio 1:3
How much does James get?

Guided

Number of shares = 1 + 3

= ....

One share is worth £120 ÷ .... = £....................

James gets ....× .... = £....................

(2 marks)

C 4 Annie and Jamil share £160 in the ratio 3:5
How much more money than Annie does Jamil get?

£.................... **(3 marks)**

C 5 Linda, Mel and Tomos share the driving on a journey in the ratio 2:3:4
Mel drove a distance of 240 km.
Work out the length of the journey.

.... km **(2 marks)**

Proportion

D **1** Mike buys 6 pencils for a total cost of £5.34
Work out the cost of 11 of these pencils.

Guided

Cost of 1 pencil = 5.34 ÷ ....
= £.....................

Cost of 11 pencils = 11 × ....
= £.....................

(2 marks)

D **2** Punita buys 3 identical notebooks for a total cost of £10.44
Work out the cost of 5 of these notebooks.

£..................... **(2 marks)**

D **3** The total cost of 4 kg of apples is £4.20
The total cost of 3 kg of apples and 2 kg of bananas is £5.05
Work out the cost of 1 kg of bananas.

First work out the cost of 1 kg of apples.

.... **(3 marks)**

C **4** A builder lays 180 bricks in 1 hour.
He always works at the same speed.
How long will it take the builder to lay 585 bricks?

Remember to include units with your answer.

.... **(2 marks)**

C **5** 4 workers can lay a stretch of road in 9 days.
How long would it take 6 workers to lay the same stretch of road?

Guided

1 worker would take 4 × .... = ... days to lay the stretch of road.

So 6 workers would take .... ÷ 6 = ... days to lay the stretch of road. **(2 marks)**

C **6** It takes one machine at a factory 24 hours to pack 12 000 boxes of cakes.
The owner of the factory buys two more machines.
All the machines work at the same rate.
How long would it take the 3 machines to pack a total of 30 000 boxes of cakes?

Work out the number of boxes that 1 machine can pack in 1 hour.

.... **(3 marks)**

Indices 2

1 Work out the value of (a) 4^{-2} (b) $49^{\frac{1}{2}}$

Guided

(a) $4^{-2} = \frac{1}{4^2}$

$= \frac{1}{\ldots\ldots}$ **(1 mark)**

(b) $49^{\frac{1}{2}} = \sqrt{49}$

$= \ldots 7 \ldots$ **(1 mark)**

B **2** Work out the value of

(a) $27^{\frac{1}{3}}$ (b) 9^{-1} (c) 4^{-3} (d) 8^0

3 .... **(1 mark)**

$\frac{1}{9}$ .... **(1 mark)**

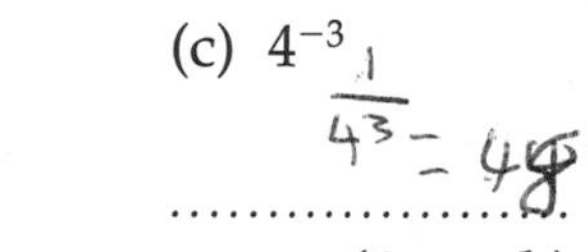

.... **(1 mark)**

1 .... **(1 mark)**

A* **3** Work out the value of (a) $8^{\frac{2}{3}}$ (b) $\left(\frac{9}{16}\right)^{-\frac{3}{2}}$

Guided

(a) $8^{\frac{2}{3}} = (8^{\frac{1}{3}})^2$

$= (\ldots\ldots)^2$

$= \ldots\ldots$ **(1 mark)**

(b) $\left(\frac{9}{16}\right)^{-\frac{3}{2}} = \left(\left(\frac{16}{9}\right)^{\frac{1}{2}}\right)^3$

$= \left(\frac{\ldots\ldots}{\ldots\ldots}\right)^3$

$= \frac{\ldots\ldots}{\ldots\ldots}$ **(2 marks)**

A* **4** Work out the value of

(a) $49^{-\frac{1}{2}}$ (b) $64^{\frac{2}{3}}$ (c) $\left(\frac{81}{16}\right)^{-\frac{3}{4}}$

.... **(1 mark)** .... **(1 mark)** .... **(2 marks)**

A* **5** Work out the value of $\frac{\sqrt{3}}{9} \times \sqrt{27}$

Write each number as a power of 3 and then use the index laws.

.... **(2 marks)**

Standard form

B **1** (a) Write 67 000 in standard form.

Exam questions similar to this have proved especially tricky – be prepared! ResultsPlus

Guided

$67\,000 = \ldots 6.7 \ldots \times 10^{4}$ **(1 mark)**

EXAM ALERT

(b) Write 2×10^{-5} as an ordinary number.

Guided

$2 \times 10^{-5} = \ldots 0.00002$ **(1 mark)**

(c) Write 760×10^{4} in standard form.

Guided

$760 \times 10^{4} = \ldots\ldots\ldots\ldots\ldots\ldots$

$= \ldots\ldots\ldots \times 10^{\ldots\ldots}$ **(1 mark)**

First write the number as an ordinary number.

B **2** (a) Write 0.54 in standard form. .... **(1 mark)**

(b) Write 7×10^{6} as an ordinary number. .... **(1 mark)**

B **3** Write these numbers in order of size.
Start with the smallest number.

Write all the numbers in standard form first.

3×10^{8} $\quad 32 \times 10^{6}$ $\quad 0.031 \times 10^{10}$ $\quad 3400 \times 10^{5}$

.... **(2 marks)**

A **4** Work out the value of $5 \times 10^{7} \times 9 \times 10^{3}$
Give your answer in standard form.

Guided

$5 \times 10^{7} \times 9 \times 10^{3} = (5 \times \ldots\ldots) \times (10^{7} \times 10^{\ldots\ldots})$

$= \ldots\ldots \times 10^{\ldots\ldots}$

$= \ldots\ldots \times 10^{\ldots\ldots}$ **(2 marks)**

A **5** Work out the value of $1.04 \times 10^{3} \div 2 \times 10^{-5}$
Give your answer in standard form.

.... **(2 marks)**

A **6** Work out the value of $7 \times 10^{5} \times 3000$
Give your answer in standard form.

.... **(2 marks)**

A **7** The number of atoms in one kilogram of helium is 1.51×10^{26}
Calculate the number of atoms in 30 kilograms of helium.
Give your answer in standard form.

.... **(2 marks)**

Calculator skills

D 1 Use your calculator to work out the value of $\frac{23.5 \times 9.4}{14.6 - 5.9}$

Write down all the figures on your calculator display.
Give your answer as a decimal.

$\frac{23.5 \times 9.4}{14.6 - 5.9} = \frac{\dots\dots\dots}{\dots\dots\dots}$

Make sure that you give your answer as a decimal. If necessary, use the S⇔D button.

= ....

(2 marks)

C 2 Use your calculator to work out the value of $\frac{\sqrt{13.5 + 3.4^2}}{2.3 \times 1.5}$

Write down all the figures on your calculator display.

.... **(3 marks)**

B 3 Use your calculator to work out the value of $\frac{45.8 \times \sin 34^\circ}{\sqrt{8.7^2 + 5.2^2}}$

Write down all the figures on your calculator display.

.... **(3 marks)**

B 4 Work out $(8.2 \times 10^{-4}) \div (3.1 \times 10^{-7})$
Give your answer in standard form correct to 3 significant figures.

.... **(2 marks)**

A 5 $y^2 = \frac{a + b}{ab}$

$a = 5 \times 10^6$ $b = 4 \times 10^3$

Find y.
Give your answer in standard form correct to 2 significant figures.

Guided

$y^2 = \frac{\dots\dots\dots + \dots\dots\dots}{5 \times 10^6 \times 4 \times 10^3}$

$y^2 = \frac{\dots\dots\dots\dots}{\dots\dots\dots\dots}$

$y = \sqrt{\dots\dots\dots\dots}$

$y = \dots\dots\dots\dots$

$y = \dots\dots\dots \times 10^{\dots}$ **(3 marks)**

Surds

1 Simplify (a) $\sqrt{48}$ (b) $\sqrt{300}$

Guided

(a) $\sqrt{48} = \sqrt{16} \times \sqrt{\ldots\ldots\ldots}$

$= \ldots\ldots\sqrt{\ldots\ldots\ldots}$

(1 mark)

(b) $\sqrt{300} = \sqrt{\ldots\ldots\ldots} \times \sqrt{\ldots\ldots\ldots}$

$= \ldots\ldots\sqrt{\ldots\ldots\ldots}$

(1 mark)

2 Rationalise the denominator of $\frac{10}{\sqrt{2}}$

Guided

$\frac{10}{\sqrt{2}} = \frac{10}{\sqrt{2}} \times \frac{\sqrt{2}}{\sqrt{2}}$

$= \frac{\ldots\ldots\sqrt{2}}{\ldots\ldots}$

$= \ldots\ldots\sqrt{2}$

(2 marks)

3 Expand and simplify $(2 - \sqrt{3})(5 + \sqrt{3})$

Use **FOIL** (**F**irst terms, **O**uter terms, **I**nner terms, **L**ast terms) to expand the brackets.

Guided

$(2 - \sqrt{3})(5 + \sqrt{3}) = 10 + 2\sqrt{3} - \ldots\ldots\ldots\sqrt{3} - \ldots\ldots\ldots$

$= \ldots\ldots\ldots - \ldots\ldots\ldots\sqrt{3}$

(2 marks)

4 Expand and simplify $(7 - \sqrt{5})(2 + \sqrt{5})$

.... **(2 marks)**

5 Expand and simplify $(3 - \sqrt{2})^2$

.... **(2 marks)**

6 Rationalise the denominator of $\frac{12 - 5\sqrt{3}}{\sqrt{3}}$

Give your answer in the form $a + b\sqrt{3}$ where a and b are integers.

.... **(3 marks)**

Problem-solving practice

D *1 Tickets R-US and Cheap Tickets both advertise tickets for the same concert.

The question has a * next to it, so make sure that you show all your working and write your answer clearly in a sentence.

Tickets R-US
£36 plus 5% booking fee

Cheap Tickets
£35 plus 7.5% booking fee

Helen wants to pay the least money possible for a ticket.
Which shop should she buy her ticket from, Tickets R-US or Cheap Tickets?

Work out the price plus the booking fee for each ticket.

(4 marks)

D 2 Last year, Kevin spent
$\frac{1}{8}$ of his salary on entertainment
$\frac{2}{5}$ of his salary on rent
15% of his salary on living expenses.
He saved the rest of his salary.
Last year Kevin's salary was £32 000
How much money did Kevin save?

Amount spent on entertainment = 32 000 ÷ ....
= £...........

You should show **all** your working.

Amount spent on rent = 32 000 ÷ .... × ....
= £...........

10% of 32 000 = ...

5% of 32 000 = ...

15% of 32 000 = ...

Amount spent on living expenses = £...........

Total amount spent = ... + ... + ...
= £...........

Amount of money saved = 32 000 − ....
= £...........

(4 marks)

Problem-solving practice

C ***3** Mr Li's garden is in the shape of a rectangle.
Part of the garden is a vegetable plot in the shape of a triangle.
The rest of the garden is grass.
Mr Li wants to spread fertiliser all over the grass.
One box of fertiliser is enough for 30 m^2 of grass.
How many boxes of fertiliser will he need?

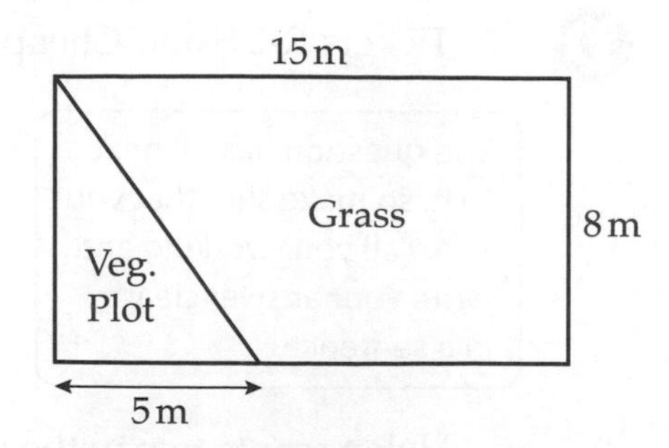

First work out the area of the grass. Divide the area by 30 to find the number of boxes – remember that Mr Li can only buy a whole number of boxes.

(4 marks)

B ***4** Kevin invests £6000 for 3 years at 4.5% simple interest in Simple Bank.
The interest is paid, by cheque, at the end of each year.
Kevin also invests £6000 for 3 years at 4.2% compound interest in Compound Bank.
Which bank pays Kevin the greater amount of total interest, Simple Bank or Compound Bank?

Simple interest means that the interest is the same each year. Work out the interest for one year then multiply this amount by 3.

For compound interest, the interest will change each year. Work out the interest for one year, add this on to the amount in the account **and** then work out the interest for the next year, and so on.

(4 marks)

5 Josip drove for 314 miles, correct to the nearest mile.
He used 34.6 litres of petrol, to the nearest tenth of a litre.

$$\text{Petrol consumption} = \frac{\text{Number of miles travelled}}{\text{Number of litres of petrol used}}$$

Work out the upper bound for the petrol consumption for Josip's journey.
Give your answer correct to 2 decimal places.

Use the upper bound for the number of miles travelled and the lower bound for the number of litres of petrol used.

.... **(3 marks)**

Algebraic expressions

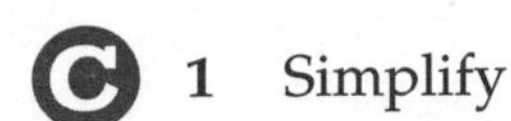

C 1 Simplify

(a) $m^3 \times m^9$ (b) $p^{10} \div p^2$ (c) $(t^4)^5$

Guided

(a) $m^3 \times m^9 = m^{3+9}$
$= m^{12}$ **(1 mark)**

(b) $p^{10} \div p^2 = p^{10-2}$
$= p^{8}$ **(1 mark)**

(c) $(t^4)^5 = t^{4 \times 5}$
$= t^{20}$ **(1 mark)**

C 2 Simplify

(a) $g \times g^6$ (b) $k^9 \div k^3 \times k^2$ (c) $(y^3)^7$

g^7 **(1 mark)** k^{12} **(1 mark)** y^{21} **(1 mark)**

C 3 Simplify

(a) $\dfrac{x^5 \times x^4}{x^6} = \dfrac{x^9}{x^6}$ (b) $\dfrac{y^{12}}{y^3 \times y}$ (c) $\left(\dfrac{z^6}{z^3}\right)^2$

x^3 **(2 marks)** .... **(2 marks)** .... **(2 marks)**

B 4 Simplify

(a) $5e^7f^2 \times 3e^4f^5$ (b) $\dfrac{28x^6y^5}{7xy^3}$ (c) $(2m^5p)^4$

Guided

(a) $5e^7f^2 \times 3e^4f^5 = 5 \times 3 \times e^{7+4} \times f^{2+5}$
$= \ldots\ldots e^{\ldots\ldots} f^{\ldots\ldots}$ **(2 marks)**

(b) $\dfrac{28x^6y^5}{7xy^3} = 28 \div 7 \times x^{6-1} \times y^{5-3}$
$= \ldots\ldots x^{\ldots\ldots} y^{\ldots\ldots}$ **(2 marks)**

(c) $(2m^5p)^4 = 2^4m^{5 \times 4}p^{1 \times 4}$
$= \ldots\ldots m^{\ldots\ldots} p^{\ldots\ldots}$ **(2 marks)**

B 5 Simplify

(a) $6cd^8 \times 4c^5d^2$ (b) $\dfrac{40a^9c^2}{8a^3c}$ (c) $(5b^3d^2)^3$

.... **(2 marks)** .... **(2 marks)** .... **(2 marks)**

A 6 Simplify

Use the index law $a^{-n} = \dfrac{1}{a^n}$

(a) $\left(\dfrac{1}{2x^3}\right)^{-2}$ (b) $\left(\dfrac{25}{64b^2c^8}\right)^{-\frac{1}{2}}$

.... **(2 marks)** .... **(2 marks)**

Arithmetic sequences

C **1** Here are the first five terms of an arithmetic sequence.

1 5 9 13 17

Find an expression, in terms of n, for the nth term of the sequence.

Guided

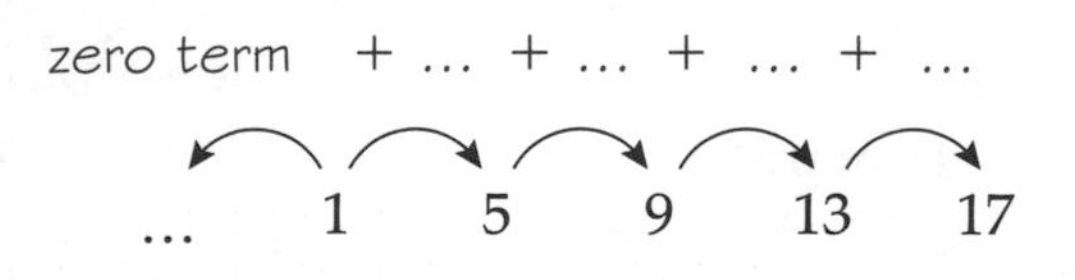

Work out the difference between each term. Then work out the zero term.

nth term = ...n + ... = ... **(2 marks)**

C **2** Here are the first five terms of an arithmetic sequence.

17 12 7 2 −3

Find an expression, in terms of n, for the nth term of the sequence.

.... **(2 marks)**

C **3** (a) Here are the first five terms of an arithmetic sequence.

3 7 11 15 19

Find an expression, in terms of n, for the nth term of the sequence.

.... **(2 marks)**

(b) Paul says that 72 is a term in this sequence.
Paul is wrong. Explain why.

Look at the type of numbers in the sequence.

....

(1 mark)

C **4** (a) The nth term of a sequence is $8n + 3$
Write down the first three terms of this sequence.

Guided

1st term $n = 1$ $8 \times 1 + 3 =$ ....

2nd term $n =$ .... $8 \times$ .... $+ 3 =$ ....

3rd term $n =$ .... $8 \times$ .... $+ 3 =$ .... **(2 marks)**

(b) Jenny says that 45 is a term in this sequence.
Jenny is wrong. Explain why.

Try and find a value for n that gives a result of 45.

.... **(1 mark)**

C **5** The nth term of a sequence is $3n - 1$
Work out the 50th term of this sequence.

.... **(1 mark)**

Expanding brackets

C **1** (a) Expand and simplify $3(2x + 5y) + 4(x - 2y)$ (b) Expand $m(2m^3 - 5)$

Guided

(a) $3(2x + 5y) + 4(x - 2y) = 6x + 15y +$ ......$x -$ ......y

$=$ ......$x +$ ......y **(2 marks)**

(b) $m(2m^3 - 5) = 2$...... $-$ ...... **(1 mark)**

C **2** Expand and simplify $7(2x - y) - 3(3x - 2y)$

Multiply each number in the second bracket by -3

.... **(2 marks)**

C **3** Expand (a) $6(3 - 4d)$ (b) $p^2(3p - 1)$

.... **(1 mark)** .... **(1 mark)**

C **4** Expand and simplify

(a) $(x + 7)(x - 3)$ (b) $(y + 6)^2$

Guided

(a) $(x + 7)(x - 3) = x^2 - 3x +$ ....$x -$ ....

$= x^2 +$.........$x -$ .... **(2 marks)**

(b) $(y + 6)^2 = (y + 6)(y + 6)$

$= y^2 + 6y +$ ....$y +$ ....

$= y^2 +$ ....$y +$ .... **(2 marks)**

C **5** Expand and simplify

(a) $(p - 4)(p - 5)$ (b) $(t - 8)^2$

.... **(2 marks)** .... **(2 marks)**

B **6** Expand and simplify $(2x + 4)(3x - 5)$

Guided

$(2x + 4)(3x - 5) =$ ....$x^2 - 10x +$ ....$x -$ ....

$=$ .... **(2 marks)**

B **7** Expand and simplify

(a) $(5x - 4)(2x - 7)$ (b) $(3y - 4)^2$ (c) $(7x + 3y)(2x - 3y)$

.... **(2 marks)** .... **(2 marks)** .... **(2 marks)**

Factorising

D **1** Factorise

(a) $6x + 18$ (b) $y^2 - 9y$

Guided (a) $6x + 18 = 6(\dots\dots\dots + \dots\dots\dots)$ **(1 mark)**

(b) $y^2 - 9y = y(\dots\dots\dots - \dots\dots\dots)$ **(1 mark)**

D **2** Factorise

(a) $5m + 20$ (b) $3v - v^2$

.... **(1 mark)** .... **(1 mark)**

EXAM ALERT C **3** Factorise fully $8mp - 12m^2$

Guided $8mp - 12m^2 = 4m(\dots\dots\dots - \dots\dots\dots)$

Take out the highest common factor.

Exam questions similar to this have proved especially tricky – be prepared! ResultsPlus

(1 mark)

C **4** Factorise fully

(a) $14ab - 21bc$ (b) $16x^2y - 40xy^2$

.... **(1 mark)** .... **(1 mark)**

B **5** Factorise

(a) $x^2 + 9x + 20$ (b) $x^2 - 25$

Guided (a) $x^2 + 9x + 20 = (x + \dots\dots\dots)(x + \dots\dots\dots)$ **(2 marks)**

(b) $x^2 - 25 = (x + \dots\dots\dots)(x - \dots\dots\dots)$ **(2 marks)**

A **6** Factorise

(a) $y^2 - 4y - 12$ (b) $y^2 - 1$

.... **(2 marks)** .... **(2 marks)**

A **7** Factorise

(a) $3x^2 + 10x + 7$ (b) $25x^2 - 9$

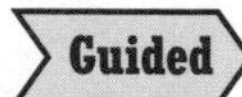

(a) $3x^2 + 10x + 7 = (3x + \dots\dots\dots)(x + \dots\dots\dots)$ **(2 marks)**

(b) $25x^2 - 9 = (5x + \dots\dots\dots)(5x - \dots\dots\dots)$ **(2 marks)**

A **8** Factorise

(a) $5x^2 + 17x - 12$ (b) $4x^2 - 49$

.... **(2 marks)** .... **(2 marks)**

A* **9** Factorise

(a) $2x^2 + 5xy + 3y^2$ (b) $8x^2 - 2xy - 3y^2$

.... **(2 marks)** .... **(2 marks)**

Linear equations 1

D **1** Solve $4(x - 3) = 16$

Guided

$4(x - 3) = 16$

$4x - \dots\dots\dots = 16$

$4x = \dots\dots\dots$

$x = \dots\dots\dots$ **(3 marks)**

D **2** Solve

(a) $7x + 5 = 33$

$x = \dots\dots\dots\dots\dots$ **(2 marks)**

(b) $3(2x + 9) = 30$

$x = \dots\dots\dots\dots\dots$ **(3 marks)**

D **3** Solve

(a) $6x - 5 = 2x + 8$

$x = \dots\dots\dots\dots\dots$ **(3 marks)**

(b) $4(x + 3) = 3x - 5$

$x = \dots\dots\dots\dots\dots$ **(3 marks)**

C **4** All the angles in the quadrilateral are measured in degrees.
Work out the size of the largest angle.

$x + 30°$, $2x$, $x - 50°$, $x + 80°$

Use the fact that angles in a quadrilateral add up to 360°.

Guided

$x + 30° + 2x + x + 80° + x - 50° = 360°$

$\dots\dots\dots x + \dots\dots\dots = 360°$

$\dots\dots\dots x = \dots\dots\dots$

$x = \dots\dots\dots$

$2x = 2 \times \dots\dots\dots = \dots\dots\dots$

$x + 80° = \dots\dots\dots\dots\dots = \dots\dots\dots$

The largest angle is $\dots\dots\dots°$ **(4 marks)**

C **5** All the lengths on the quadrilateral are measured in centimetres.
The perimeter of the quadrilateral is 39 cm.
Work out the length of the shortest side of the quadrilateral.

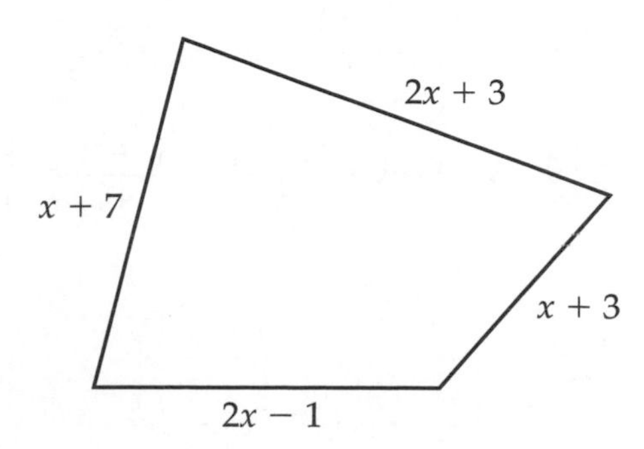

$\dots\dots\dots\dots\dots$ cm **(4 marks)**

Linear equations 2

B 1 Solve $\frac{6x + 5}{4} = x - 2$

Guided

EXAM ALERT

$\cancel{4} \times \frac{(6x + 5)}{\cancel{4}} = 4(x - 2)$

$6x + 5 = \ldots\ldots x - \ldots\ldots$

$\ldots\ldots x + 5 = - \ldots\ldots$

$\ldots\ldots x = - \ldots\ldots$

$x = - \frac{\ldots\ldots}{\ldots\ldots}$

Exam questions similar to this have proved especially tricky – be prepared! ResultsPlus

(3 marks)

B 2 Solve $\frac{6 - 2x}{3} = 1 - x$

$x = \ldots\ldots\ldots\ldots$ **(3 marks)**

B 3 Solve $\frac{5x}{3} - 1 = x + 5$

Guided

$\frac{5x}{3} - 1 = x + 5$

$\frac{5x}{\cancel{3}} \times \cancel{3} - 1 \times 3 = x \times 3 + 5 \times 3$

$5x - \ldots\ldots = 3x + \ldots\ldots$

$\ldots\ldots x = \ldots\ldots$

$x = \ldots\ldots$

(4 marks)

B 4 Solve $\frac{x}{4} - 2 = 2(x - 3)$

$x = \ldots\ldots\ldots\ldots$ **(4 marks)**

A 5 Solve $\frac{2x - 1}{3} + \frac{x + 3}{4} = \frac{5}{6}$

Multiply each term by the LCM of 3, 4 and 6.

$x = \ldots\ldots\ldots\ldots$ **(5 marks)**

Straight-line graphs

D 1 On the grid draw the graph of $y = 2x + 5$ for values of x from -3 to 2

First draw a table of values.
Next work out the values of y.

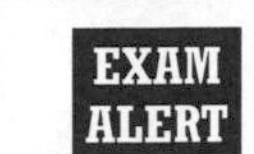

x	-3	-2	-1	0	1	2
y						

$x = -3$: $y = (2 \times -3) + 5 =$ ....

Exam questions similar to this have proved especially tricky – be prepared! ResultsPlus

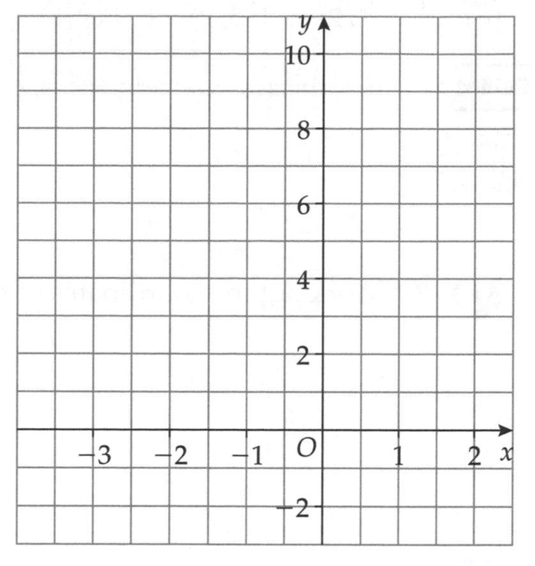

Plot your points on the grid. Join your points with a straight line using a ruler.

(3 marks)

C 2 A straight line has equation $y = 3 - 2x$

Write down (a) the gradient and (b) the coordinates of the y-intercept of this straight line.

Use $y = mx + c$ where m is the gradient and c is the y-intercept.

(a) .... **(1 mark)**

(b) .... **(1 mark)**

B 3 Find the equation of

(a) line A

Guided

gradient $= \frac{.........}{.........}$

$=$ ....

y-intercept $= (0, ...)$

Equation of line A is $y = ...x + ...$

(2 marks)

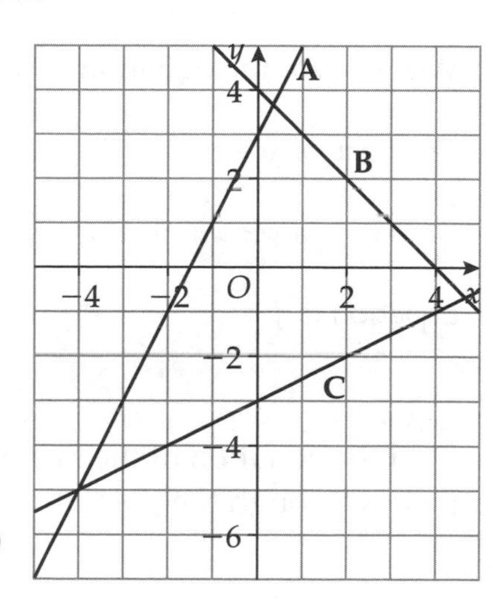

(b) line B

$y =$ .... **(2 marks)**

(c) line C

$y =$ .... **(2 marks)**

A 4 A straight line passes through the points (0, 3) and (4, 5).

Find the equation of the straight line.

$y =$ .... **(3 marks)**

Parallel and perpendicular

D 1 Work out the coordinates of the midpoint of the line joining (1, 2) and (5, 9).

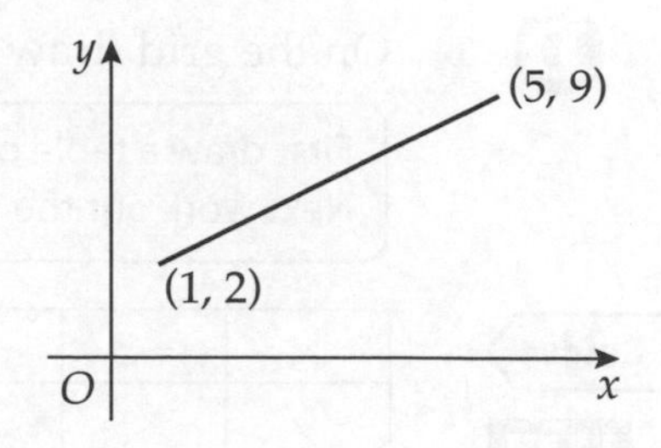

Guided

Coordinates of midpoint $= \left(\frac{1 + 5}{2}, \frac{\ldots\ldots + \ldots\ldots}{2}\right)$

$= (\ldots\ldots\ldots, \ldots\ldots\ldots)$ **(2 marks)**

D 2 Work out the coordinates of the midpoint of the line joining $(-3, 1)$ and $(7, 5)$.

.... **(3 marks)**

B 3 Write down the equation of a line parallel to the line with equation $y = 4x - 5$

Use $y = mx + c$ where m is the gradient and c is the y-intercept. Remember that parallel lines have the same gradient.

.... **(1 mark)**

B 4 Write down the equation of a line perpendicular to the line with equation $y = 2x + 6$

Guided

Gradient of $y = 2x + 6$ is ....

Gradient of perpendicular line is $-\frac{1}{\ldots\ldots}$

Equation is $y = -\frac{1}{\ldots\ldots}x + \ldots\ldots\ldots$

(1 mark)

A 5 Find the equation of the line which is parallel to the line with equation $y = 3x - 2$ and which passes through the point (0, 4).

.... **(2 marks)**

A 6 A straight line, L, passes through the point with coordinates (4, 1) and is perpendicular to the line with equation $y = 2x - 5$
Find the equation of the line L.

.... **(3 marks)**

3-D coordinates

C 1 A cuboid is shown on a 3-D grid.

Coordinates are always given in the order (x, y, z).

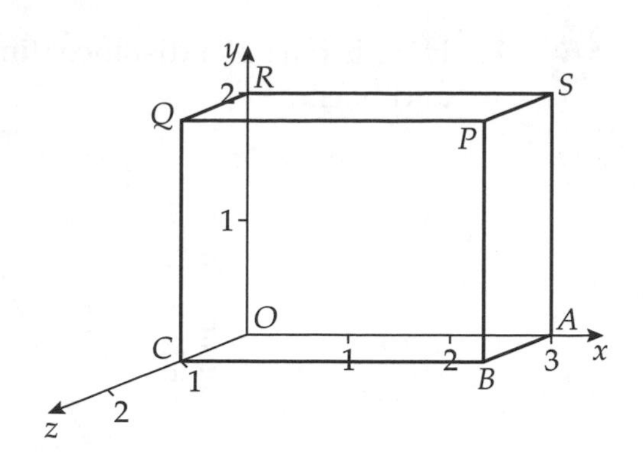

(a) Write down the letter of the point with coordinates (3, 0, 1).

.... **(1 mark)**

(b) Write down the coordinates of the point P.

(......... , ... , ...) **(1 mark)**

C 2 A cuboid is drawn on a 3-D grid.
The point Q has coordinates (5, 2, 3).
Write down the coordinates of point P.

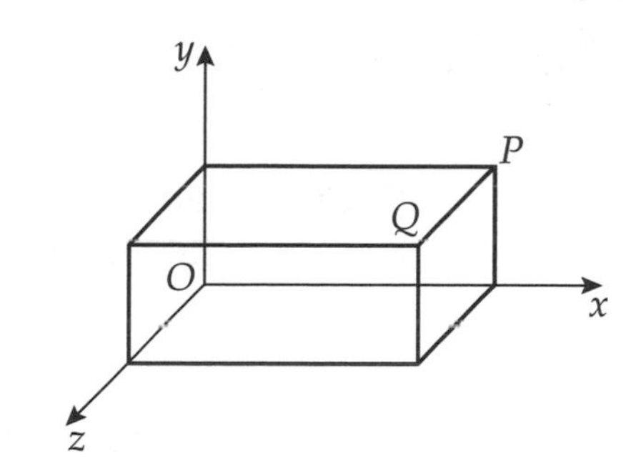

(......... , ... , ...) **(2 marks)**

B 3 F and G are two points on a 3-D coordinate grid.
Point F has coordinates (4, 1, 5).
Point G has coordinates (8, −3, −4).
Work out the coordinates of the midpoint of the line segment FG.

To find the coordinates of the midpoint, find the mean of the x-coordinates, the mean of the y-coordinates and the mean of the z-coordinates.

Coordinates of midpoint $= \left(\frac{4 + 8}{2}, \frac{\text{.........} + \text{.........}}{2}, \frac{\text{.........} + \text{.........}}{2}\right)$

= (........... , ... , ...) **(2 marks)**

4 The diagram shows a cuboid on a 3-D grid.
P, Q, R and S are vertices of the cuboid.

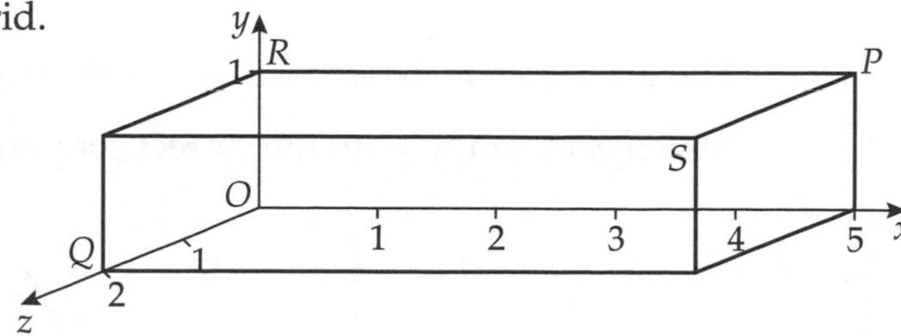

(a) Write down the coordinates of S.

(......... , ... , ...) **(1 mark)**

(b) Work out the coordinates of the midpoint of the line segment PQ.

(......... , ... , ...) **(2 marks)**

Real-life graphs

C 1 Here is part of a distance–time graph showing Kelly's journey from her house to the gym and back.

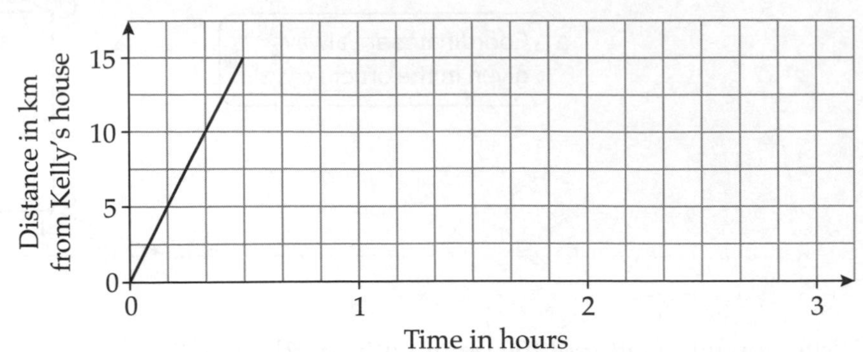

(a) Work out Kelly's speed for the first part of her journey. Give your answer in km/h.

Guided

Speed = $\frac{\text{distance}}{\text{time}}$

Give the time in hours.

= $\frac{\ldots\ldots\ldots}{\ldots\ldots\ldots}$ = ... km/h **(2 marks)**

(b) She spends 90 minutes at the gym and then travels home at a speed of 45 km/hour. Complete the distance–time graph.

Guided

90 minutes = ... hours

Time taken to get home = $\frac{\text{distance}}{\text{speed}}$

= $\frac{15}{\ldots\ldots\ldots}$

= .... hour

Draw a horizontal line on the graph to show the time she spent at the gym.

Draw a line on the graph to show Kelly's journey home. **(2 marks)**

B 2 Water is poured into each container at a constant rate.

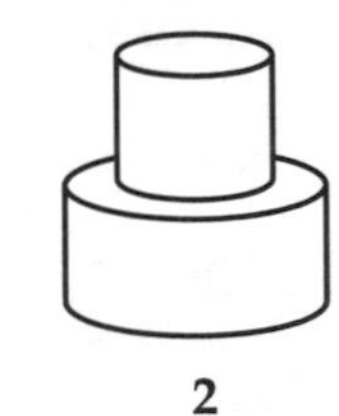

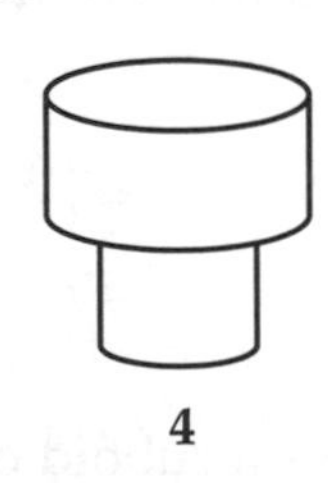

1 2 3 4

The graphs show how the depth of the water in each container changes with time.

Match each graph with the correct container.

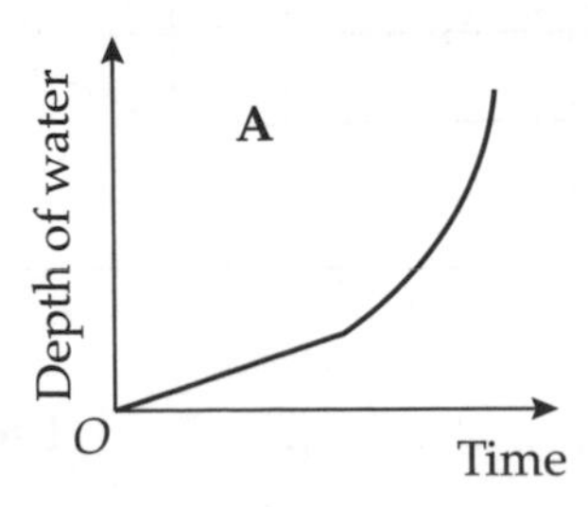

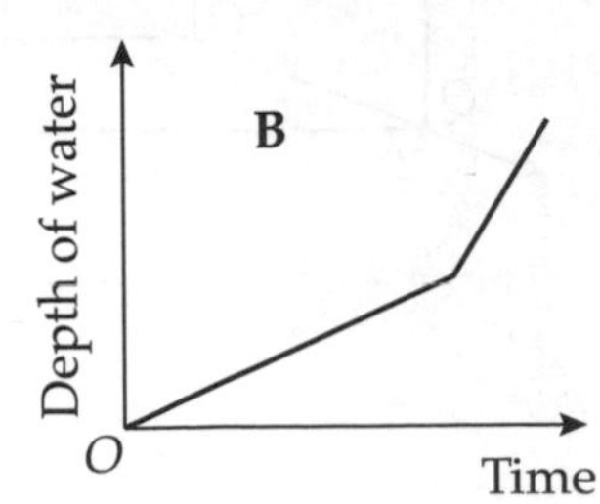

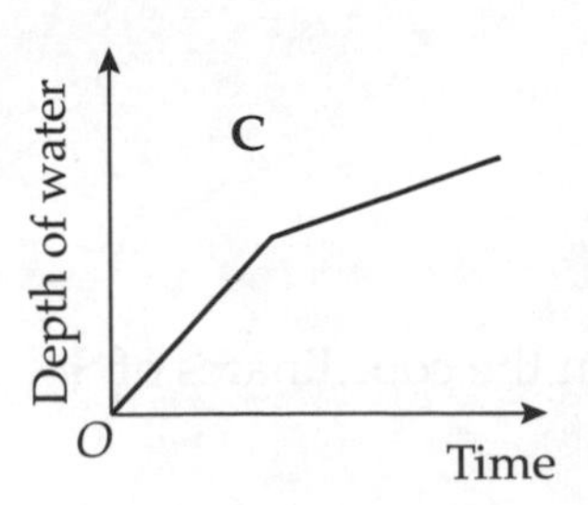

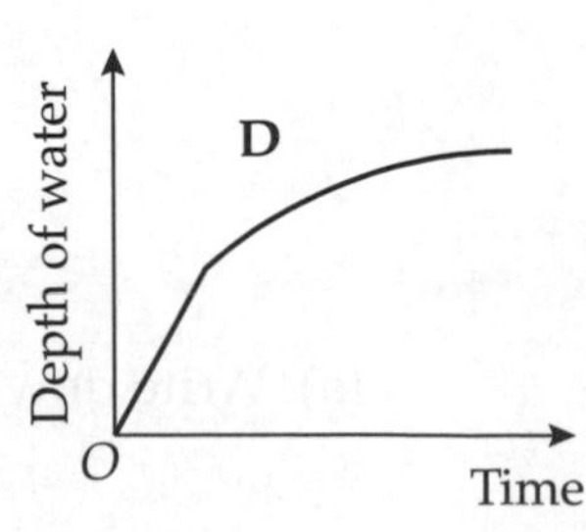

A and .... B and .... C and .... D and ....

(2 marks)

Formulae

D **1** $A = 3t^2 - bc$
Work out the value of A when $b = 3$, $c = 6$ and $t = -5$

Guided $A = 3 \times (-5)^2 - (3) \times (6)$
$= 3 \times$ .... − ....
$=$ .... − ....
$=$ ....

(2 marks)

D **2** Find the value of $4x^2 - 5x$ when $x = -3$

.... **(2 marks)**

D **3** Lesley hires a car.
The initial cost of hiring the car is £45, plus £30 for each day.
Write down a formula for the total cost, £C, to hire the car for t days.

Guided Total cost = 45 + 30 × number of days
.... = 45 + 30.........

Replace the words by the letters given in the question.

(2 marks)

C **4** A shop sells cakes and buns.
Cakes cost c pence each, buns cost b pence each.
Malik buys 6 cakes and 4 buns.
The total cost is C pence.
Write down a formula for C in terms of c and b.

.... **(3 marks)**

C **5** The diagram shows a quadrilateral.
Write down a formula, in terms of x, for the perimeter, P, of the quadrilateral.
Give your answer in its simplest form.

$2x + 1$
$2x - 3$
$x + 4$
$3x - 7$

.... **(3 marks)**

C **6** A bag of apples costs a pence.
A bag of pears costs p pence.
Sue buys 4 bags of apples and 3 bags of pears.
The total cost is £C.
Write down a formula for C in terms of a and p.

.... **(3 marks)**

Rearranging formulae

1 Make t the subject of the formula
$v = u + 6t$

$v = u + 6t$

$v - \ldots\ldots\ldots = 6t$

$\dfrac{v - \ldots\ldots\ldots}{\ldots\ldots\ldots} = t$

$t = \dfrac{v - \ldots\ldots\ldots}{\ldots\ldots\ldots}$

(2 marks)

2 Make k the subject of the formula
$p = 2(k - 3)$

$k = \ldots\ldots\ldots\ldots\ldots\ldots\ldots$ **(3 marks)**

3 Make h the subject of the formula
$M = \sqrt{\dfrac{2h + 1}{3}}$

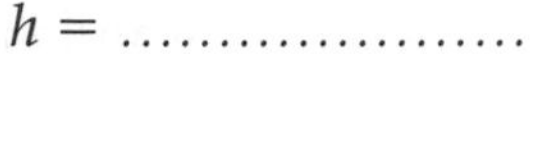

$h = \ldots\ldots\ldots\ldots\ldots\ldots\ldots$ **(3 marks)**

4 Make y the subject of
$3(y + 2) = a(5 - 2y)$

$3(y + 2) = a(5 - 2y)$

$3y + \ldots\ldots\ldots = \ldots\ldots\ldots - 2ay$

$3y + 2ay = \ldots\ldots\ldots - \ldots\ldots\ldots$

$y(\ldots\ldots\ldots + \ldots\ldots\ldots) = \ldots\ldots\ldots - \ldots\ldots\ldots$

$y = \dfrac{\ldots\ldots\ldots - \ldots\ldots\ldots}{\ldots\ldots\ldots + \ldots\ldots\ldots}$

(4 marks)

5 Make p the subject of the formula

$t = \sqrt{\dfrac{2p - 4}{p}}$

$p = \ldots\ldots\ldots\ldots\ldots\ldots\ldots$ **(4 marks)**

Inequalities

 **1** (a) $-3 < n \leqslant 2$
n is an integer.
Write down all the possible values of n.

Guided n could be -2, ..., ..., ..., 2 **(2 marks)**

(b) Solve the inequality $4x - 3 > 21$

Exam questions similar to this have proved especially tricky – be prepared! ResultsPlus

Guided **EXAM ALERT**

$4x - 3 > 21$

$4x >$ ....

$x >$ .... **(2 marks)**

B **2** (a) An inequality is shown on the number line.

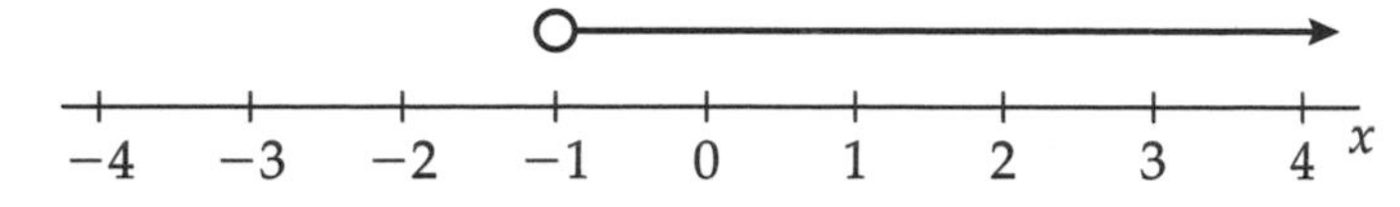

Write down the inequality.

.... **(2 marks)**

(b) Solve the inequality $5x + 1 \leqslant 3x - 17$

.... **(2 marks)**

 **3** (a) Solve the inequality $2 - 4y > 12 - y$

Guided

$2 > 12$ .... Add $4y$ to each side.

.... $>$ ....y Subtract 12 from each side.

.... $> y$ **(2 marks)**

(b) y is an integer.
Write down the largest value of y that satisfies $2 - 4y > 12 - y$

$y =$ .... **(1 mark)**

A **4** Solve $-2 < \frac{3x}{5} < 7$

Solve the two equations $-2 < \frac{3x}{5}$ and $\frac{3x}{5} < 7$

.... **(3 marks)**

Inequalities on graphs

B **1** The region **R** satisfies the inequalities

$x \geqslant 1, y \geqslant 2, x + y \leqslant 5$

On the grid below, draw straight lines and use shading to show the region **R**.

Guided

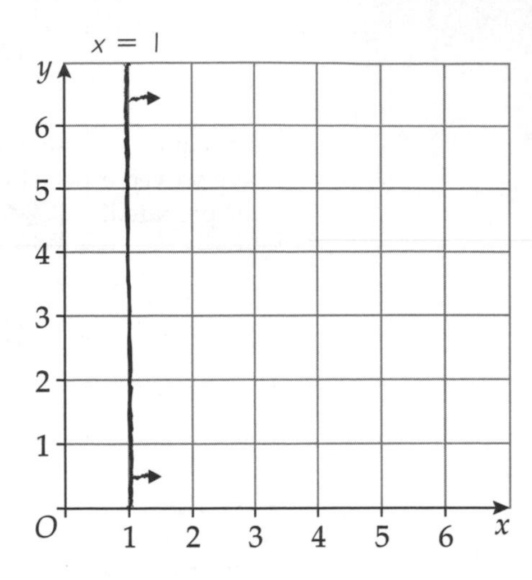

1. Draw the line $y = 2$ on the grid.
   Use a small arrow to show which side of this line you want.
2. Next find some points on the line $x + y = 5$.
3. Use your points to draw the line $x + y = 5$ on the grid.
   Use a small arrow to show which side of this line you want.
4. Shade in the region you want and label it **R**.

$y = 5 - x$

x	0	2	4
y	5	………	………

(4 marks)

A **2** $x > 0, y > -1, y < 2x, 5x + 4y < 20$

x and y are both integers.

On the grid, mark with a cross (×) each of the **seven** points which satisfy all **four** inequalities.

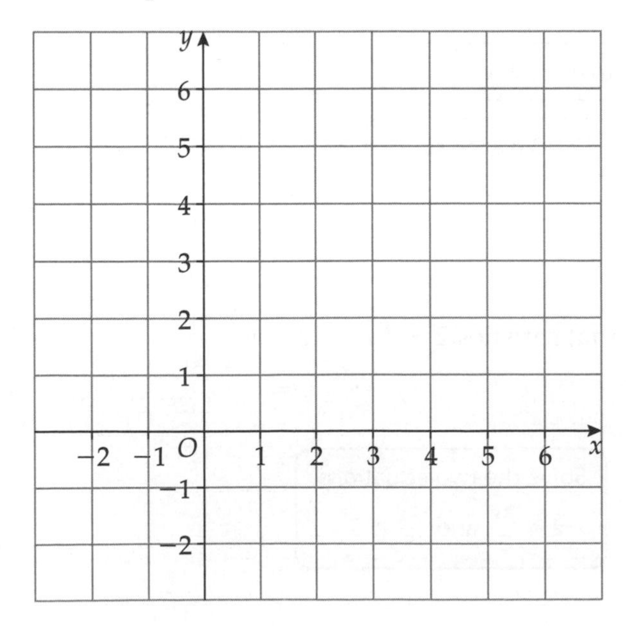

(4 marks)

Quadratic and cubic graphs

1 (a) Complete the table of values for $y = x^2 + 2x - 5$

x	-2	-1	0	1	2	3
y		-6	-5			10

Write all the values in the table.

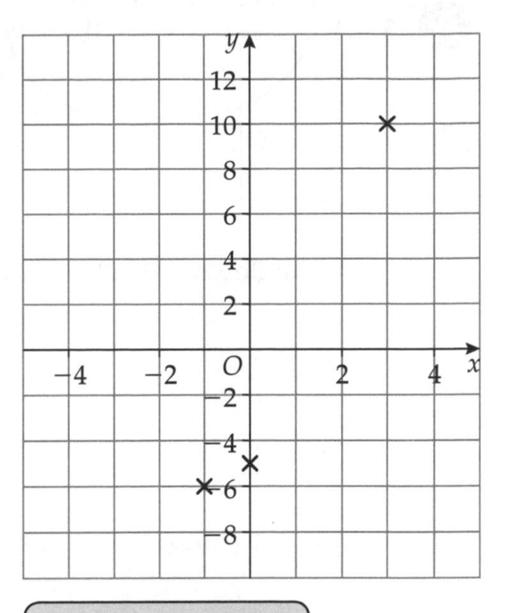

Guided

$x = -2 \quad y = (-2)^2 + 2 \times -2 - 5$

$= \ldots\ldots\ldots$

$x = 1 \quad y = 1^2 + 2 \times 1 - 5$

$= \ldots\ldots\ldots$

$x = 2 \quad y = 2^2 + 2 \times 2 - 5$

$= \ldots\ldots\ldots$

(2 marks)

(b) On the grid, draw the graph of $y = x^2 + 2x - 5$

Plot the points from the table and join with a smooth curve.

(2 marks)

A 2 (a) Complete the table of values for $y = x^3 - 3x - 1$

x	-2	-1	0	1	2	3
y		1	-1			17

(2 marks)

(b) On the grid, draw the graph of $y = x^3 - 3x - 1$ **(2 marks)**

(c) Use your graph to find estimates for the values of x when $x^3 - 3x - 1 = 0$

Find the values of x where the graph crosses the x-axis.

....

(d) Use your graph to find estimates of the solutions to the equation $x^3 - 3x - 1 = -2$

Draw the line $y = -2$ on the graph. Find the x-coordinates of the points of intersection with the curve.

....

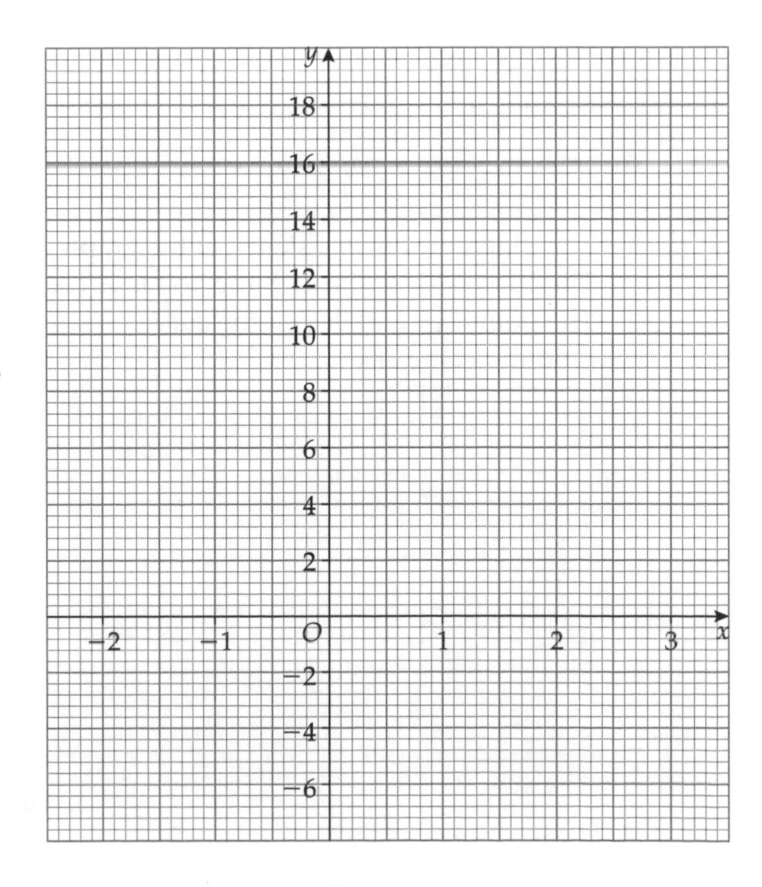

(2 marks)

Graphs of $\frac{k}{x}$ and a^x

1

A
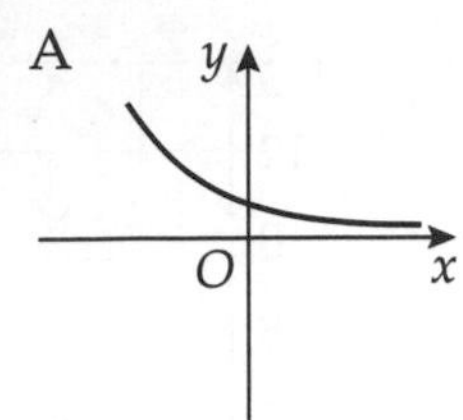

B
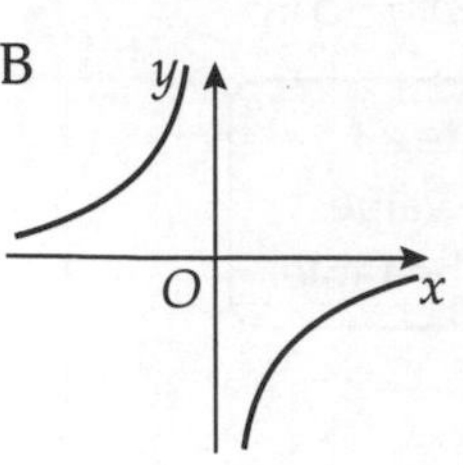

C
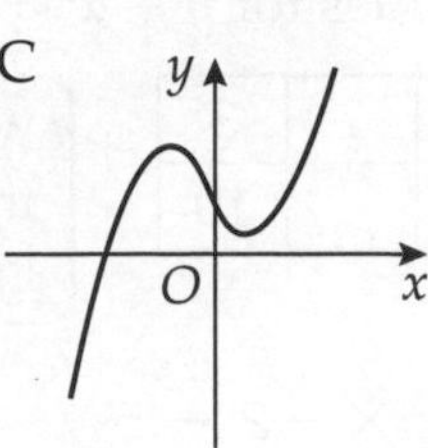

D E
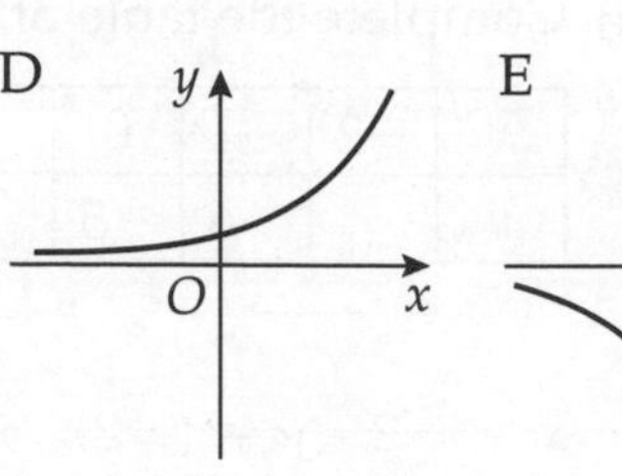

Write down the letter of the graph that could have equation

(a) $y = 3^x$ ....

(b) $y = -\frac{3}{x}$ ....

(c) $y = x^3 + 4x^2 - 2x + 3$ .... **(2 marks)**

2 The sketch shows a curve with equation $y = ka^x$ where k and a are constants, and $a > 0$
The curve passes through the points (1, 2) and (3, 32).
Calculate the value of k and the value of a.

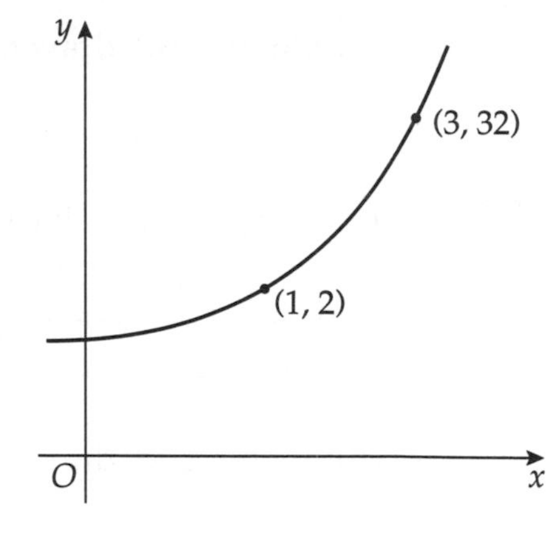

Using (3, 32): $32 = ka^3$ (1)

Using (1, 2): $2 = ka^1$ (2)

(1) ÷ (2) ...... $= a^2$

$a = \sqrt{.........}$

$a =$ ......

Using (2): $2 = k$......

$k =$ ....

(3 marks)

3 The diagram shows a sketch of the graph $y = ab^x$.
The curve passes through the points $A(\frac{1}{3}, \frac{1}{2})$ and $B(2, 16)$.
The point $C(-\frac{1}{3}, k)$ lies on the curve.
Find the value of k.

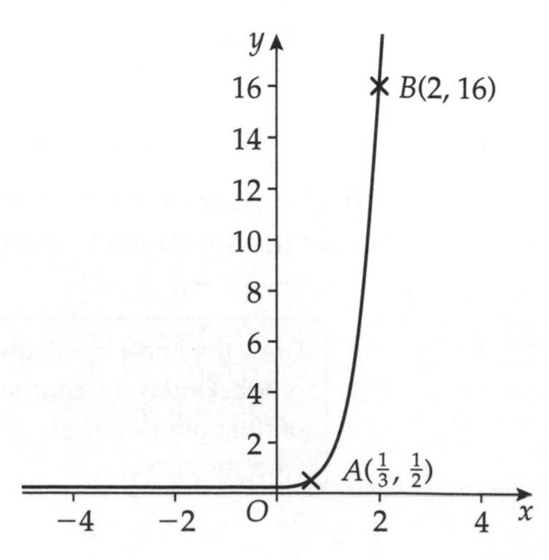

$k =$ .... **(4 marks)**

Trial and improvement

1 The equation $x^3 + 3x = 21$ has a solution between 2 and 3
Use a trial and improvement method to find this solution.
Give your answer correct to 1 decimal place.
You must show **all** your working.

x	$x^3 + 3x$	Too big or too small
2.5	23.125	too big
2.4	....	....
2.3	....	....
2.35	....	....

$x =$ .... **(4 marks)**

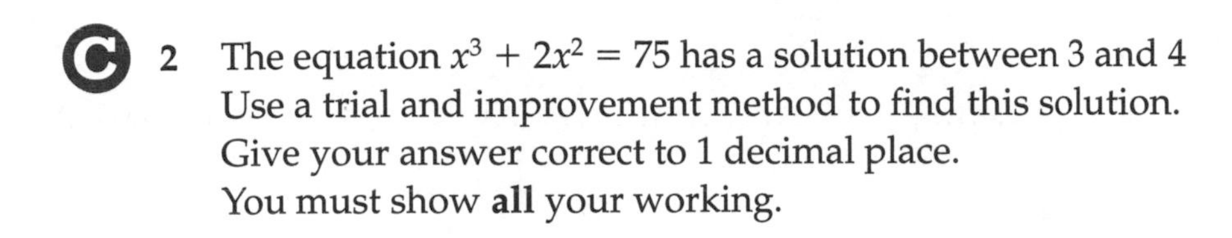

2 The equation $x^3 + 2x^2 = 75$ has a solution between 3 and 4
Use a trial and improvement method to find this solution.
Give your answer correct to 1 decimal place.
You must show **all** your working.

$x =$ .... **(4 marks)**

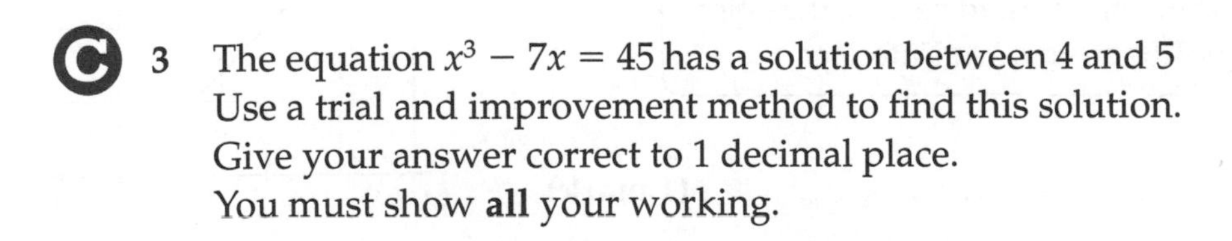

3 The equation $x^3 - 7x = 45$ has a solution between 4 and 5
Use a trial and improvement method to find this solution.
Give your answer correct to 1 decimal place.
You must show **all** your working.

$x =$ .... **(4 marks)**

Simultaneous equations 1

1 Solve the simultaneous equations
$x + 3y = 5$
$4x + 5y = 6$

Exam questions similar to this have proved especially tricky – be prepared! ResultsPlus

Guided

EXAM ALERT

$x + 3y = 5$ (1)
$4x + 5y = 6$ (2)
.........x + ...y = ... (1) × 4
− $4x + 5y = 6$ (2)
.... y = ...
y = ...
Substitute y = ... into (1):
$x + 3 \times$ .... $= 5$
$x +$ .... $= 5$
x = ...
Solution is x = ..., y = ... **(3 marks)**

2 Solve the simultaneous equations
$2x - 5y = -6$
$4x + 3y = 1$

x = ..., y = ... **(3 marks)**

3 The graphs of the straight lines with equations
$x + y = 4$ and $2y = x + 5$ have been drawn on the grid.
Use the graph to solve the simultaneous equations
$x + y = 4$
$2y = x + 5$

Write down the coordinates of the point where the straight lines intersect.

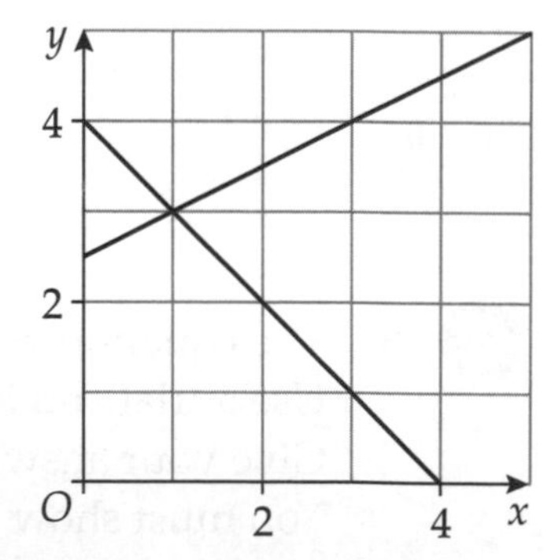

x = ..., y = ... **(1 mark)**

A* **4** Three apples and two bananas have a total cost of £1.55
Four apples and three bananas have a total cost of £2.20
Work out the cost of one apple and the cost of one banana.

Set up a pair of simultaneous equations. Use a for apple and b for banana. Change the prices from pounds to pence.

.... **(5 marks)**

Quadratic equations

 **1** Solve

Exam questions similar to this have proved especially tricky – be prepared! ResultsPlus

(a) $x^2 + 8x = 0$

$x^2 + 8x = 0$

$x(x + \ldots 8 \ldots) = 0$

$x = 0$ or $x + \ldots 8 \ldots = 0$

$x = \ldots -8 \ldots$

(2 marks)

(b) $x^2 - 6x - 27 = 0$

$x^2 - 6x - 27 = 0$

$(x + \ldots\ldots)(x - \ldots\ldots) = 0$

$x + \ldots\ldots = 0$ or $x - \ldots\ldots = 0$

$x = \ldots\ldots$ or $x = \ldots\ldots$

(3 marks)

 **2** Solve

(a) $x^2 - 7x + 10 = 0$

$x = \ldots\ldots\ldots\ldots$ **(3 marks)**

(b) $x^2 + 10x + 9 = 0$

$x = \ldots\ldots\ldots\ldots$ **(3 marks)**

A **3** (a) Solve $2x^2 + 13x + 21 = 0$

Guided

$2x^2 + 13x + 21 = 0$

$(2x + \ldots\ldots)(x + \ldots\ldots) = 0$

$2x + \ldots\ldots = 0$ or $x + \ldots\ldots = 0$

$x = \ldots\ldots$ or $x = \ldots\ldots$

(3 marks)

(b) Solve $9x^2 - 25 = 0$

Use the difference of two squares.

$x = \ldots\ldots\ldots\ldots$ **(2 marks)**

 **4** The diagram shows a trapezium.
All measurements on the diagram are in centimetres.
The area of the trapezium is 72 cm^2.

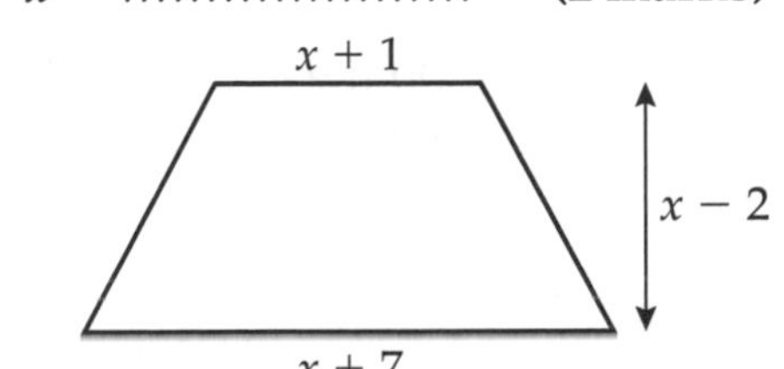

(a) Show that $x^2 + 2x - 80 = 0$

Use information from the diagram to form an equation for the area. Then rearrange the equation to the form given.

(3 marks)

(b) Work out the lengths of the parallel sides of the trapezium.

Solve the quadratic equation from (a) and use the positive value of x to find the lengths of the parallel sides.

$\ldots\ldots\ldots\ldots$ cm

$\ldots\ldots\ldots\ldots$ cm **(4 marks)**

 **5** Solve $6x(x + 2) = 5 - x$

Expand the brackets then gather all the x terms onto the left-hand side of the equation.

$x = \ldots\ldots\ldots\ldots$ **(4 marks)**

Completing the square

1 For all values of x,
$x^2 - 8x + 19 = (x - p)^2 + q$
Find the value of p and the value of q.

Guided

$x^2 - 8x + 19 = (x - \text{.........})^2 - \text{.........}^2 + 19$

$= (x - \text{.........})^2 + \text{.........}$

$p = \text{.........}$, $q = \text{.........}$ **(2 marks)**

2 For all values of x,
$x^2 + 12x - 5 = (x + a)^2 + b$
Find the value of a and the value of b.

$a = \text{.....................}$, $b = \text{.....................}$ **(2 marks)**

3 (a) For all values of x,
$x^2 - 2x + 7 = (x - p)^2 + q$
Find the value of p and the value of q.

$p = \text{.....................}$, $q = \text{.....................}$ **(2 marks)**

(b) On the axes, sketch the graph of $y = x^2 - 2x + 7$

The minimum point of the graph will be (p, q).

Substitute $x = 0$ into the equation to find where the graph crosses the y-axis and mark this on the graph.

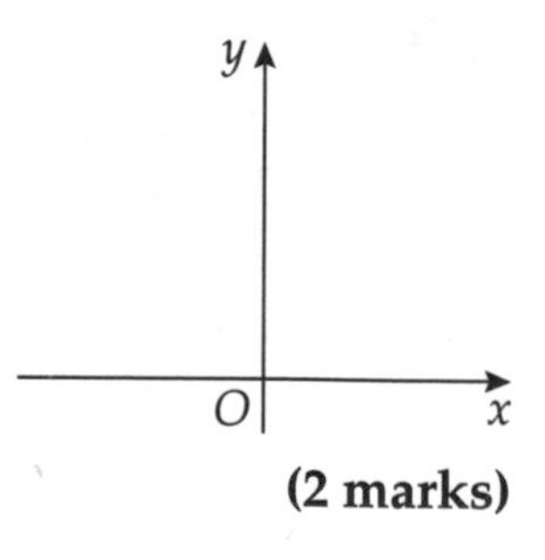

(2 marks)

4 Given that $x^2 - 6x + a = (x + b)^2$ for all values of x, find the value of a and the value of b.

Start by expanding $(x + b)^2$.

$a = \text{.....................}$, $b = \text{.....................}$ **(3 marks)**

5 Here is a sketch of the graph $y = x^2 + 8x + 9$
Find the coordinates of P and the coordinates of Q.

Start by completing the square for the equation.

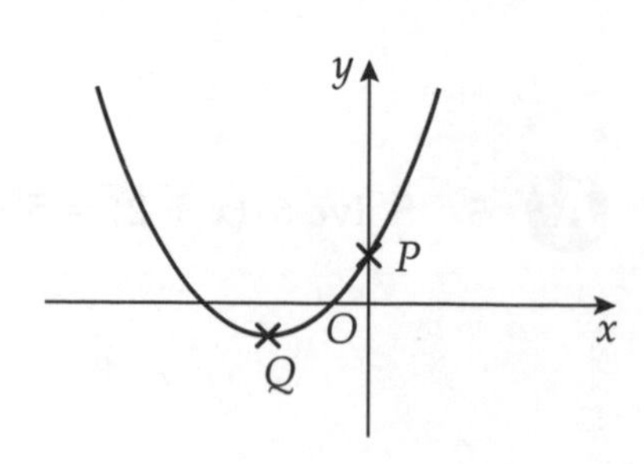

P ...., Q .... **(4 marks)**

The quadratic formula

1 Solve $x^2 + 3x - 9 = 0$
Give your solutions correct to 3 significant figures.

$$x = \frac{-b \pm \sqrt{b^2 - 4ac}}{2a}$$

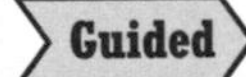

$a = 1$, $b = \ldots\ldots$, $c = -9$

$$x = \frac{-\ldots\ldots \pm \sqrt{(\ldots\ldots)^2 - 4 \times \ldots\ldots \times \ldots\ldots}}{2 \times \ldots\ldots}$$

$$x = \frac{-\ldots\ldots \pm \sqrt{\ldots\ldots}}{\ldots\ldots}$$

$x = \ldots\ldots\ldots\ldots$ or $\ldots\ldots\ldots\ldots$

$x = \ldots\ldots\ldots\ldots$ or $\ldots\ldots\ldots\ldots$ correct to 3 s.f. **(3 marks)**

2 Solve $5x^2 - 2x - 8 = 0$
Give your solutions correct to 3 significant figures.

$x = \ldots\ldots\ldots\ldots$ **(3 marks)**

3 The diagram shows a 6-sided shape.
All the corners are right angles.
All the measurements are given in centimetres.
The area of the shape is 80 cm².

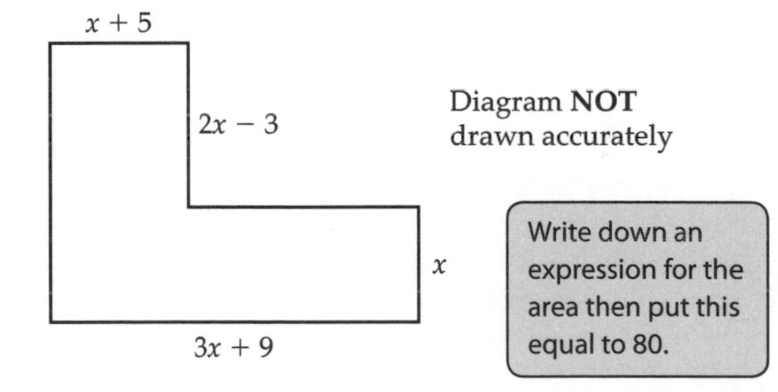

(a) Show that $5x^2 + 16x - 95 = 0$

(3 marks)

(b) Work out the length of the longest side of the 6-sided shape.

Solve the equation from (a). Use the value of x to find the value of $2x + 3$.

$\ldots\ldots\ldots\ldots$ cm **(4 marks)**

Quadratics and fractions

 **1** Solve $\frac{x}{2x-5} - \frac{4}{x-1} = 1$

Guided

$\frac{x\cancel{(2x-5)}(x-1)}{\cancel{(2x-5)}} - \frac{4(2x-5)\cancel{(x-1)}}{\cancel{(x-1)}} = 1(2x-5)(x-1)$

$x(x-1) - 4(2x-5) = (2x-5)(x-1)$

Expand the brackets.

Rearrange to the form $ax^2 + bx + c = 0$.

Solve the quadratic equation by factorising.

$x =$ .... **(5 marks)**

 **2** Solve $\frac{9}{2x-1} = \frac{x+7}{x+1}$

$x =$ .... **(5 marks)**

 **3** Solve $\frac{3}{x-2} + \frac{x}{x+1} = 2$

Give your solutions correct to 3 significant figures.

$x =$ .... **(5 marks)**

 **4** Solve $\frac{1}{x^2-16} + \frac{2}{x+4} = \frac{3x}{x-4}$

Start by factorising $x^2 - 16$.

$x =$ .... **(5 marks)**

Equation of a circle

1 (a) On the grid, construct the graph of $x^2 + y^2 = 16$

The graph of $x^2 + y^2 = 16$ is a circle of radius $\sqrt{16}$.

(1 mark)

(b) Solve the simultaneous equations
$x^2 + y^2 = 16$
$x + y = 2$

Draw the graph of $x + y = 2$ and find the points of intersection of the circle and the straight line.

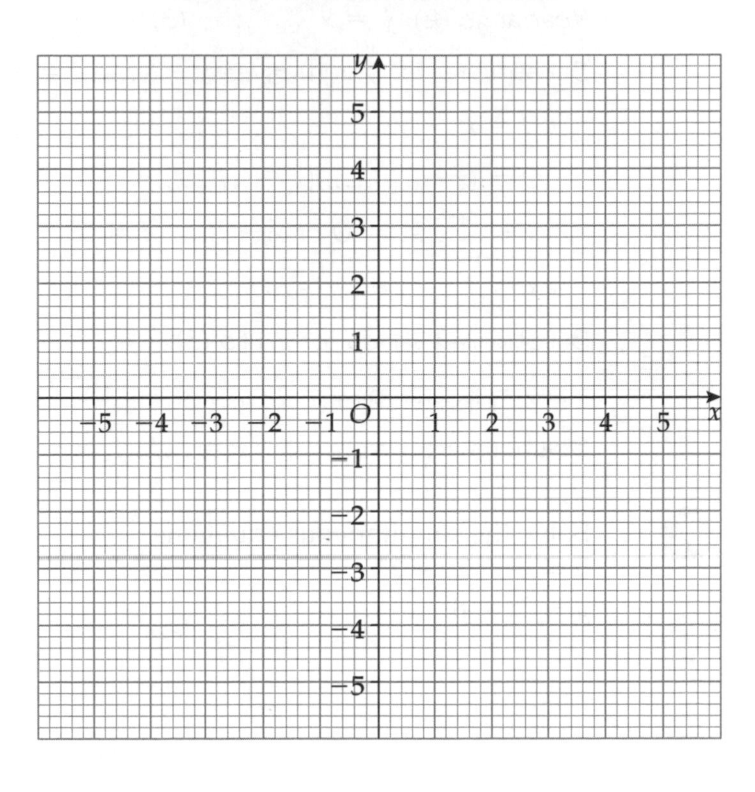

$x =$ ...., $y =$ .... **(3 marks)**

2 The diagram shows a circle of radius 2, centre the origin.
Solve the equations
$x^2 + y^2 = 4$
$2y - 1 = x$

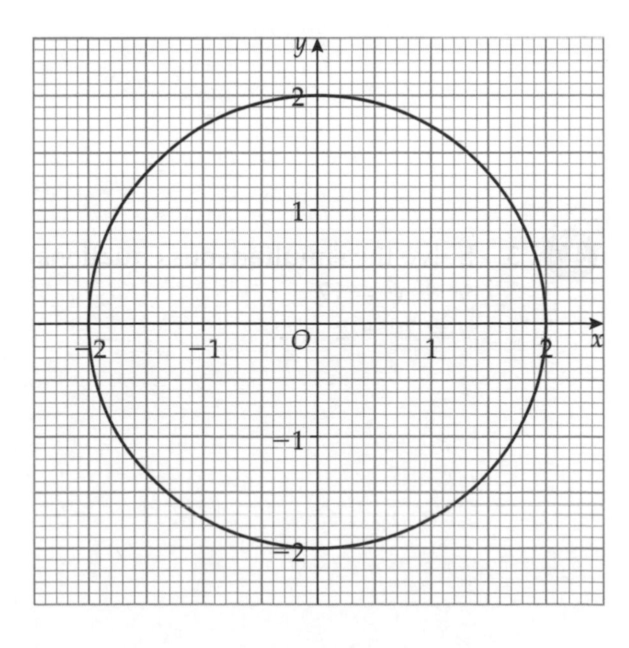

$x =$ ...., $y =$ .... **(3 marks)**

Simultaneous equations 2

1 Solve the simultaneous equations
$y = x^2 - 3x - 16$
$y + 4 = x$

Guided

$y = x^2 - 3x - 16$ (1)

$y + 4 = x$ (2)

Rearrange (2): $y = x$ .... (3)

Substituting for y in (1) gives $x -$ .... $= x^2 - 3x - 16$

$0 = x^2 -$ ....$x -$ ....

$0 = (x +$ ....$)(x -$ ....$)$

$x +$ .... $= 0$ or $x -$ .... $= 0$

$x =$ .... or $x =$ ....

Substituting values for x into (3):

$y =$ .... or $y =$ ....

Solutions are $x =$ ...., $y =$ .... and $x =$ .... and $y =$ .... **(5 marks)**

2 Solve the simultaneous equations
$y = x^2 + 2x - 22$
$y - 3x = 8$

$x =$ ...., $y =$ .... **(5 marks)**

3 Solve the simultaneous equations
$y^2 + x^2 = 17$
$y = 2x + 2$

Start by substituting $2x + 2$ for y in $y^2 + x^2 = 17$.

$x =$ ...., $y =$ .... **(6 marks)**

Direct proportion

A **1** In a spring the tension, T newtons, is directly proportional to its extension, x cm. When the tension is 200 newtons, the extension is 8 cm.

(a) Find a formula for T in terms of x.

Guided

$T \propto x$ so $T = kx$

Using $T = 200$, $x = 8$ 200 $= k$ 8

$k = \frac{200}{8}$

$k =$ 25

Formula is $T =$ 25 x **(3 marks)**

(b) Calculate the tension, in newtons, when the extension is 16 cm.

Substitute $x = 16$ into your formula to work out the value of T.

$T =$ .... newtons **(1 mark)**

A **2** The volume of liquid in a tube, $V\text{ cm}^3$, is directly proportional to the height, h cm, of the liquid.
When the height of the liquid is 6 cm, the volume of liquid is 4 cm^3.
Work out the height of the liquid when the volume is 2.5 cm^3.

$h =$ .... **(4 marks)**

A **3** The power, P watts, of an engine is proportional to the square of the speed, s m/s, of the engine.
When $s = 20$, $P = 640$

(a) Find a formula for P in terms of s.

Use the relationship $P \propto s^2$.

$P =$ .... watts **(3 marks)**

(b) Find the value of s when $P = 360$

$s =$ .... **(1 mark)**

A **4** B is directly proportional to l^3.
When $l = 4$, $B = 24$
Work out the value of B when $l = \frac{2}{3}$

$B =$ .... **(4 marks)**

Proportionality formulae

 1 m is inversely proportional to the square of t.
When $t = 4$, $m = \frac{1}{2}$

(a) Find a formula for m in terms of t.

$m \propto \frac{1}{t^2}$ so $m = \frac{k}{t^2}$

Guided

Using $t = 4$, $m = \frac{1}{2}$ .... $= \frac{k}{.........^2}$

$k =$ .... × ....

$k =$ ....

Formula is $m = \frac{.........}{t^2}$ **(3 marks)**

(b) Calculate a value of t when $m = 4.5$

Substitute $m = 4.5$ into your formula to work out a value for t.

$t =$ .... **(2 marks)**

 2 l is inversely proportional to h.
When $h = 6$, $l = 40$
Find the value of h when $l = \frac{1}{2}$

Go through the same steps as in question 1.

$h =$ .... **(5 marks)**

 3 The shutter speed of a camera, S, is inversely proportional to the square of the aperture setting, f.
When $f = 10$, $S = 80$
Find the value of S when $f = 16$

$S =$ .... **(4 marks)**

 4 The energy, E joules, of an object varies directly with the square of its speed, s m/s.
When $s = 0.6$, $E = 54$
Work out the speed when $E = 1350$

$s =$ .... m/s **(5 marks)**

Transformations 1

A* **1** The graph of $y = f(x)$ is shown on the grids below.

(a) On this grid, sketch the graph of $y = f(x) + 2$

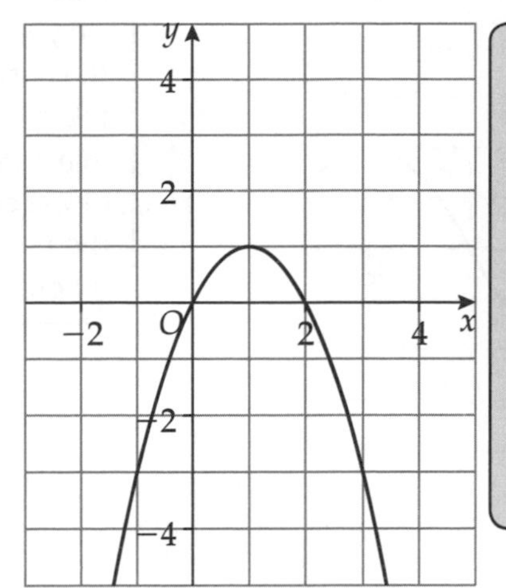

The graph goes through (0, 0), (2, 0) and (1, 1). Draw the images of these points and then sketch in the rest of the graph.

(2 marks)

(b) On this grid, sketch the graph of $y = f(\frac{1}{2}x)$.

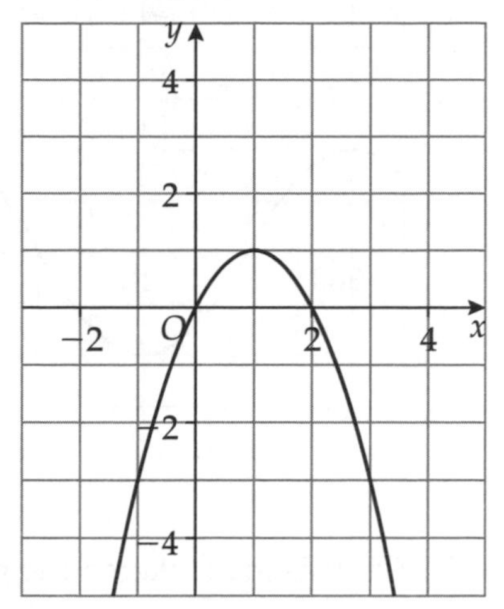

(2 marks)

A* **2** The diagram shows part of the curve with equation $y = f(x)$.
The coordinates of the maximum point of this curve are (3, 4).
Write down the coordinates of the maximum point of the curve with equation

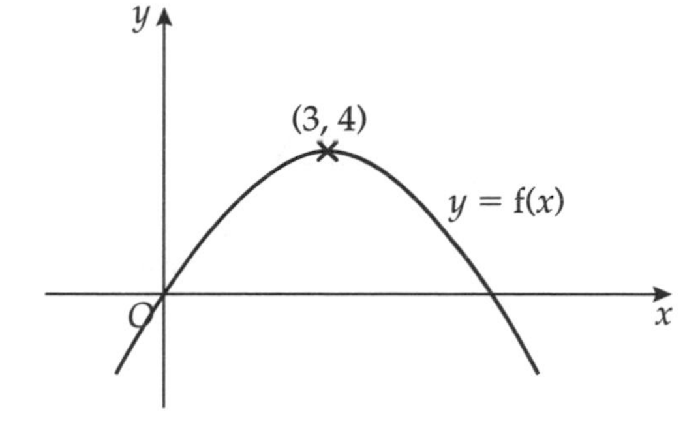

(a) $y - f(x - 5)$ .... **(1 mark)**

(b) $y = f(-x)$ .... **(1 mark)**

(c) $y = 3f(x)$ .... **(1 mark)**

A* **3** The graph of $y = f(x)$ is shown on the grids below.

(a) On this grid, sketch the graph of $y = -f(x)$.

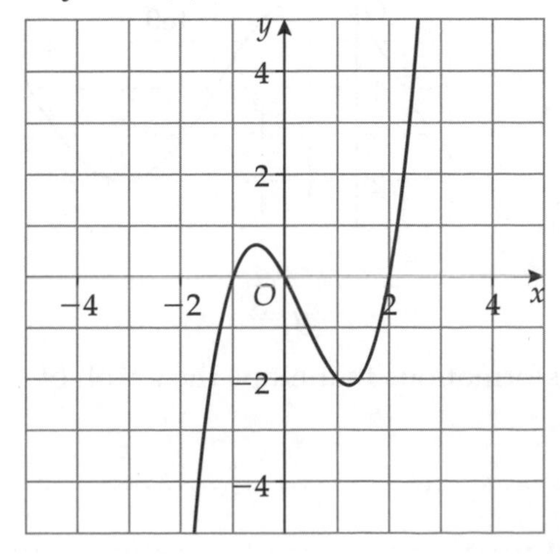

(2 marks)

(b) On this grid, sketch the graph of $y = f(x + 3)$.

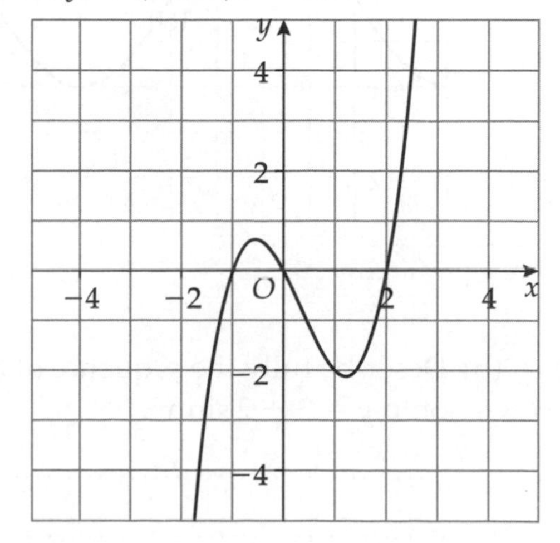

(2 marks)

Transformations 2

1 The graph of $y = \cos x$ is drawn on the grids below.

(a) On this grid, sketch the graph of $y = \cos(x + 90)°$.

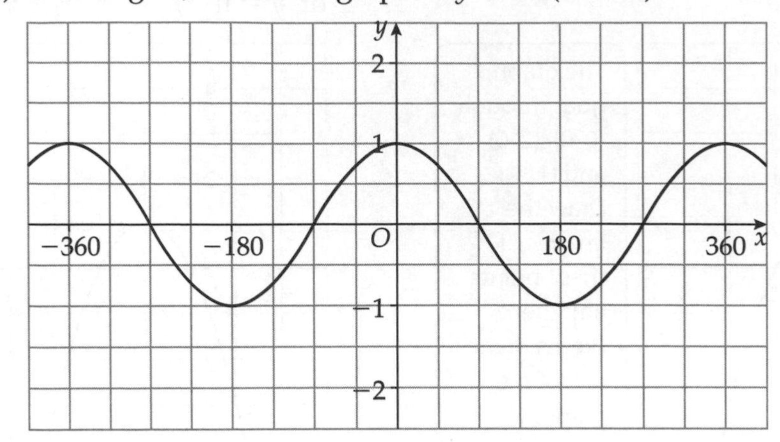

On your graph mark the positions of the maximum and minimum points of $y = \cos(x + 90)°$ then complete the graph.

(2 marks)

(b) On this grid, sketch the graph of $y = 2\cos x°$.

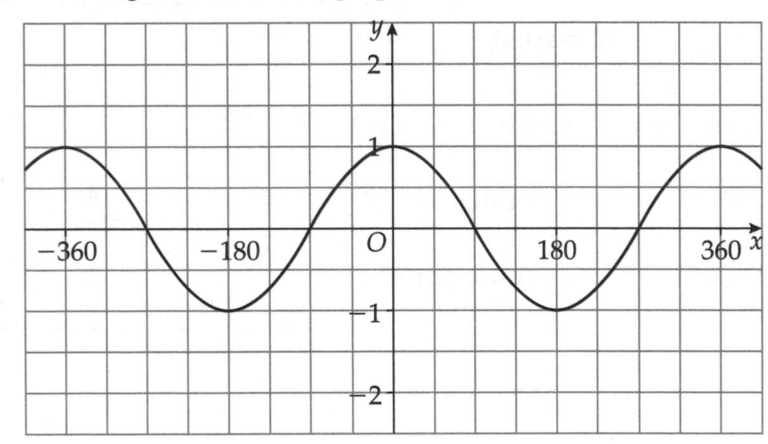

(2 marks)

2 Graph A shows the graph of $y = \sin x°$.

(a) Use this to help you write down the equation of graph B.

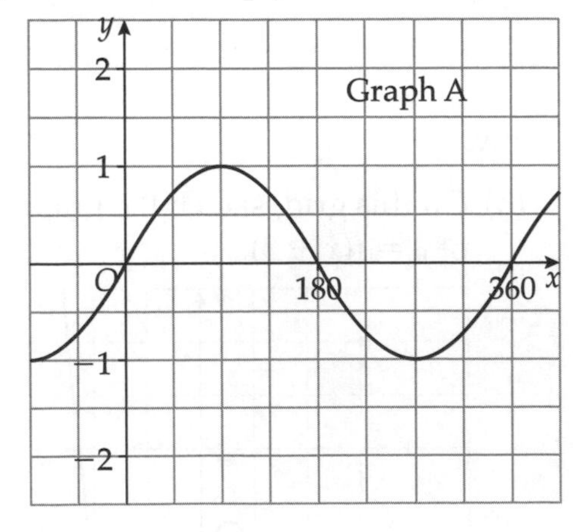

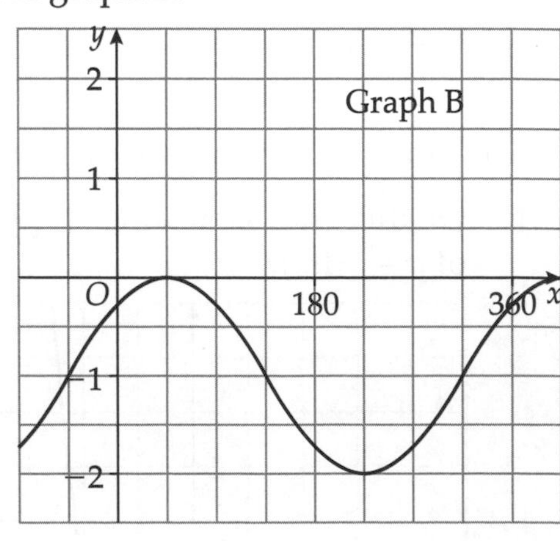

$y =$ .... **(2 marks)**

(b) Describe fully the sequence of two transformations that maps the graph of $y = \sin x°$ onto $y = 3 + 2\sin x°$.

....

....

(2 marks)

Algebraic fractions

1 Simplify fully

Exam questions similar to this have proved especially tricky – be prepared!

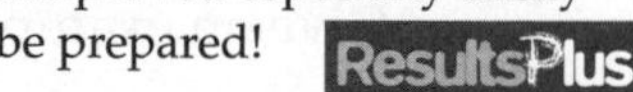

(a) $\frac{5x+15}{x^2+3x}$

(b) $\frac{x^2-25}{x^2-2x-35}$

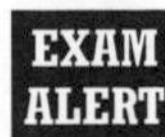

Guided

(a) $\frac{5x+15}{x^2+3x} = \frac{5(\ldots\ldots + \ldots\ldots)}{x(\ldots\ldots + \ldots\ldots)}$

$= \frac{\ldots\ldots\ldots\ldots}{\ldots\ldots\ldots\ldots}$ **(3 marks)**

(b) $\frac{x^2-25}{x^2-2x-35} = \frac{(x+\ldots\ldots)(x-\ldots\ldots)}{(x+\ldots\ldots)(x-\ldots\ldots)}$

$= \frac{\ldots\ldots\ldots\ldots}{\ldots\ldots\ldots\ldots}$ **(3 marks)**

2 Simplify fully

(a) $\frac{x-4}{2x^2-8x}$

.... **(2 marks)**

(b) $\frac{2x^2-17x+21}{4x^2-9}$

.... **(3 marks)**

3 Write as a single fraction in its simplest form

(a) $\frac{2x-1}{5} + \frac{x+2}{3}$

.... **(3 marks)**

(b) $\frac{5x+1}{2} - \frac{4-x}{5}$

.... **(3 marks)**

4 Write as a single fraction in its simplest form

$\frac{3}{2x+1} + \frac{8}{2x^2-7x-4}$

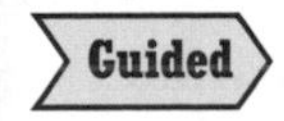

Guided

$\frac{3}{2x+1} + \frac{8}{2x^2-7x-4} = \frac{3}{2x+1} + \frac{8}{(2x+1)(x-\ldots\ldots)}$

$= \frac{3(x-\ldots\ldots)}{(2x+1)(x-\ldots\ldots)} + \frac{8}{(2x+1)(x-\ldots\ldots)}$

$= \frac{3x-\ldots\ldots+8}{(2x+1)(x-\ldots\ldots)}$

$= \frac{\ldots\ldots\ldots\ldots\ldots\ldots}{(2x+1)(x-\ldots\ldots)}$

(3 marks)

5 Write as a single fraction in its simplest form

$\frac{4}{x^2+8x+15} - \frac{1}{2x^2+5x-3}$

.... **(4 marks)**

Proof

 **1** Chloe says, 'For any whole number, n, the value of $5n - 1$ is always a square number.'
Chloe is **wrong**.
Give an example to show that she is wrong.

Guided

Let $n = 1$: $5n - 1 = 5 \times 1 - 1 = 4 \rightarrow$ square number

Let $n = 2$: $5n - 1 = 5 \times$ .... $- 1 =$ .... $\rightarrow$ ....

Continue trying different values for n until the result is not a square number.

(1 mark)

 **2** Ben says, 'The product of any two prime numbers is always odd.'
He is **wrong**.
Explain why.

(2 marks)

 **3** Prove that the sum of any four consecutive even numbers is always a multiple of 4

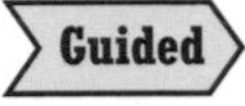

Let n be an integer then $2n$ will be an even number.

1st even number $= 2n$

2nd even number $= 2n + 2$

3rd even number $=$ ....

4th even number $=$ ....

Sum of four consecutive even numbers $= 2n + 2n + 2 +$ .... $+$ ....

$=$ ....$n +$ ....

$= 4($.........$n +$ ....$)$

As 4 is a factor, the sum of any four consecutive even numbers is always a multiple of 4.

(3 marks)

 **4** Prove that the sum of any three consecutive odd numbers is always a multiple of 3

(3 marks)

 **5** Prove that $(4n + 1)^2 - (4n - 1)^2$ is always a multiple of 8, for all positive integer values of n.

(3 marks)

Problem-solving practice

D 1 Here is a triangle.

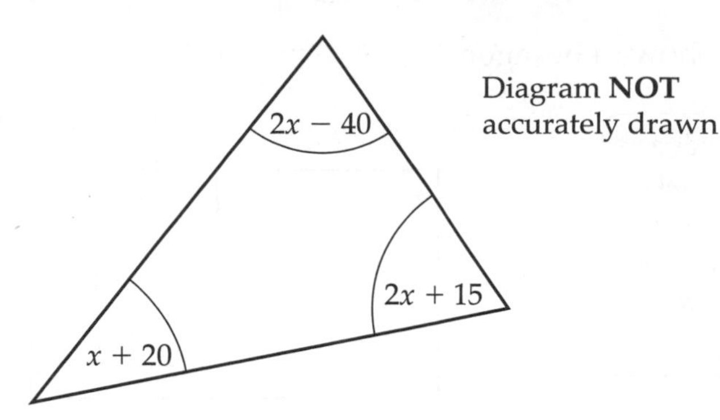

Diagram **NOT** accurately drawn

All angles are measured in degrees.
Work out the size of the largest angle.

Guided

$x + 20 + 2x + 15 + \ldots\ldots\ldots\ldots - \ldots\ldots\ldots\ldots = 180$

$\ldots\ldots\ldots\ldots x - \ldots\ldots\ldots\ldots = 180$

$\ldots\ldots\ldots\ldots x = 180 + \ldots\ldots\ldots\ldots$

$\ldots\ldots\ldots\ldots x = \ldots\ldots\ldots\ldots$

$x = \ldots\ldots\ldots\ldots$

Largest angle: $2x + 15 = 2 \times \ldots\ldots\ldots\ldots \ldots \ldots\ldots\ldots\ldots$

$= \ldots\ldots\ldots\ldots^\circ$

Use the fact that angles in a triangle add up to 180°.

(4 marks)

D 2 On the grid, draw the graph of $y = 3x + 2$ for values of x from -1 to 3

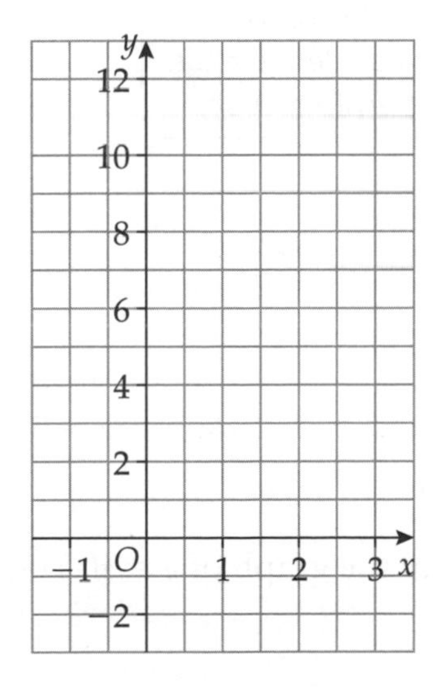

Start by drawing a table of values for x from -1 to 3.
Use the equation $y = 3x + 2$ to work out the corresponding values of y.
Plot your points and then join them with a straight line.

(3 marks)

B 3 AB is a line segment.
A is the point $(2, 5, -6)$.
The midpoint M of the line segment AB has coordinates $(4, -1, -2)$.
Jacob says that B has coordinates $(3, 2, -4)$.
Jacob is wrong. Explain why.

Work out what happens to each coordinate as you move from A to M. Then repeat these increases or decreases to move from M to B.

(2 marks)

Problem-solving practice

*4 The diagram shows a hexagon.

The question has a * next to it, so make sure that you show all your working and write your answer clearly.

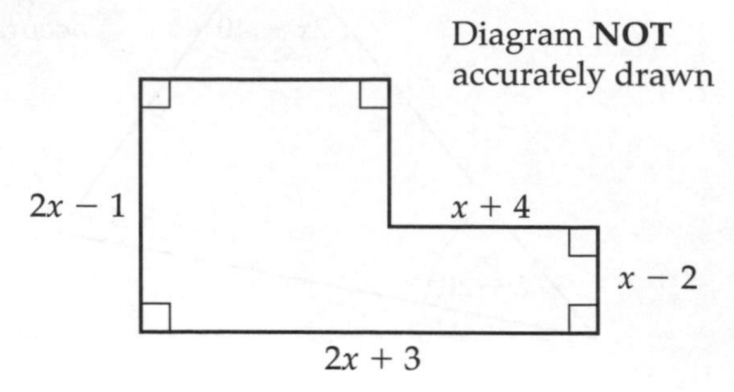

All measurements are in cm.
Show that the area of the hexagon can be written as $3x^2 - x - 7$

Divide the shape into two rectangles and find an expression for any missing lengths. Write down an expression for the area of each rectangle. Add your expressions together and then simplify the answer.

(4 marks)

5

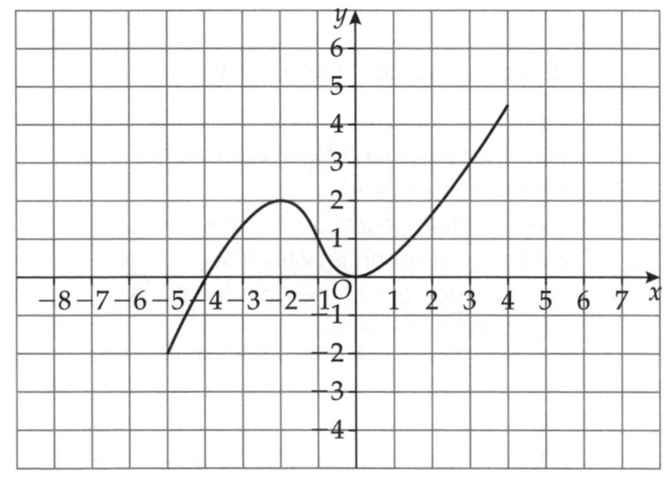

Sketch the graph of $y = f(x - 3)$ on the grid and then describe the transformation.

The graph of $y = f(x)$ is shown on the grid.

(a) Fully describe the single transformation that would map the graph of $y = f(x)$ onto the graph of $y = f(x - 3)$

....

....

(2 marks)

(b) Fully describe the single transformation that would map the graph of $y = f(x)$ onto the graph of $y = -f(x)$

....

....

(2 marks)

Angle properties

D **1** *ABCD* is a straight line.
PQ is parallel to *RS*.

(a) Write down the size of the angle marked x.
Give a reason for your answer.

(b) Work out the size of the angle marked y.
Give a reason for your answer.

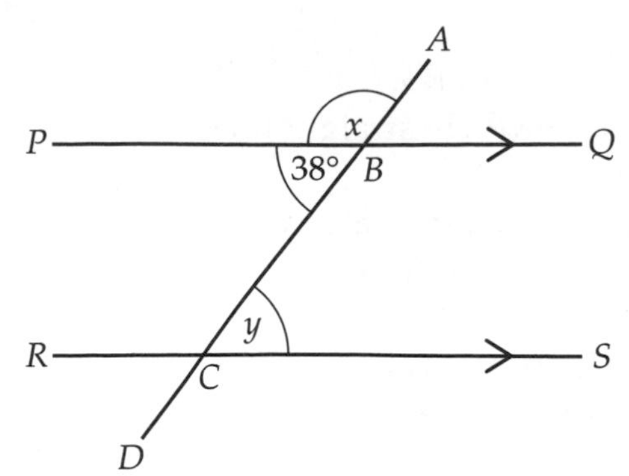

Diagram **NOT** drawn accurately

Guided

(a) $x =$ ……..°

Angles on a straight line add up to ……..° **(2 marks)**

(b) $y =$ ……..°

…………………… angles are equal. **(2 marks)**

D **2** *AB* is parallel to *CD*.
EF is a straight line.
Write down the value of y.
Give a reason for your answer.

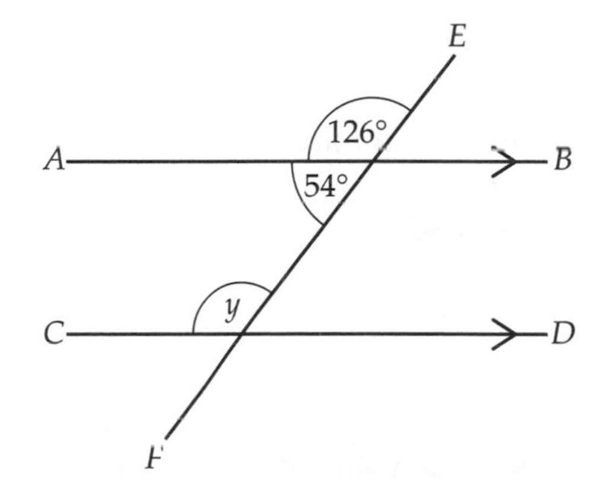

Diagram **NOT** drawn accurately

…………………° **(2 marks)**

C **3** *ABC* is parallel to *DEFG*.
$BF = EF$.
Angle $CBF = 48°$.
Work out the size of angle *BEF*.
You must give reasons for your answer.

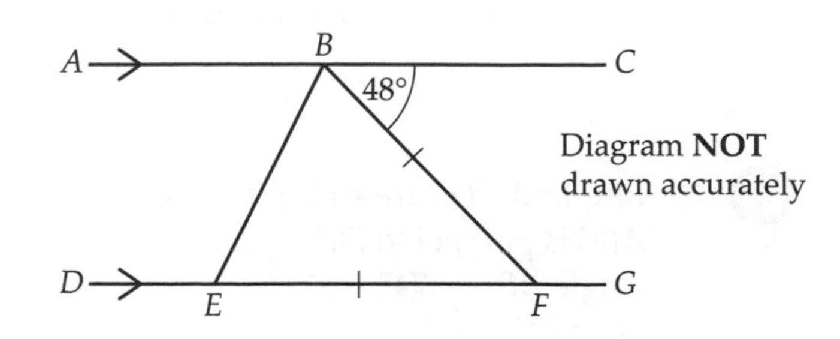

Diagram **NOT** drawn accurately

…………………° **(3 marks)**

Solving angle problems

D **1** PQ is parallel to RS.
OSQ and ORP are straight lines.

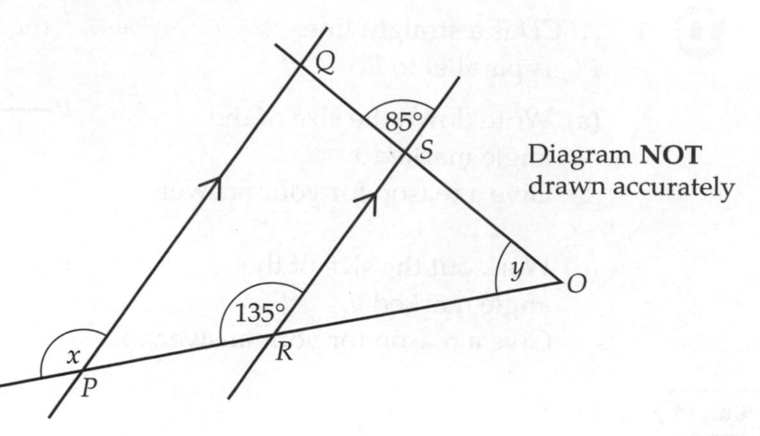

(a) Write down the value of x.
Give a reason for your answer.

(b) Work out the value of y.
Give reasons for your answer.

Guided

(a) $x =$ ....°

.... angles are equal. **(2 marks)**

(b) Angle $SRO = 180° - 135°$ — Angles on a straight line add up to ....°

$=$ ....°

Angle $RSO =$ ....° — .... angles are equal.

$y = 180° -$ .... $-$ .... — Angles in a triangle add up to ....°

$=$ ....° **(3 marks)**

D **2** AC and BD are straight lines which cross at E.
AD is parallel to BC.

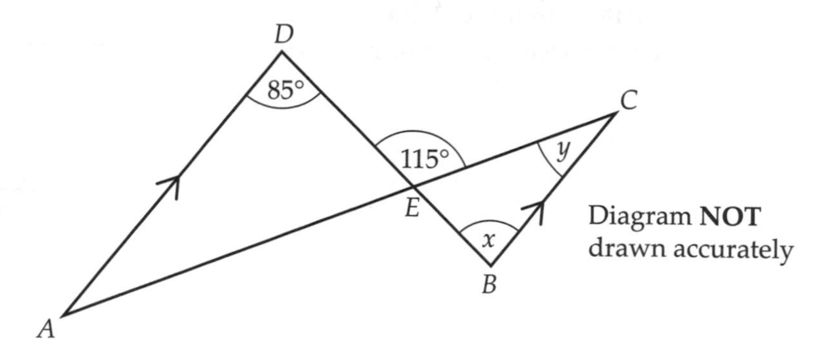

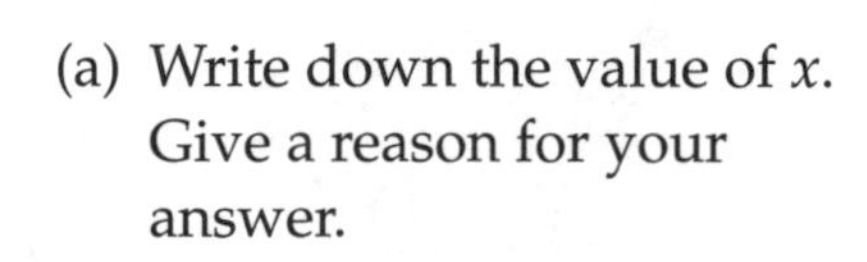

(a) Write down the value of x.
Give a reason for your answer.

...................° **(2 marks)**

(b) Work out the value of y.
Give reasons for your answer.

...................° **(3 marks)**

C **3** BEG and CFG are straight lines.
ABC is parallel to DEF.
Angle $ABE = 74°$.
$EF = EG$.
Work out the size of the angle marked y.
You must give reasons for your answer.

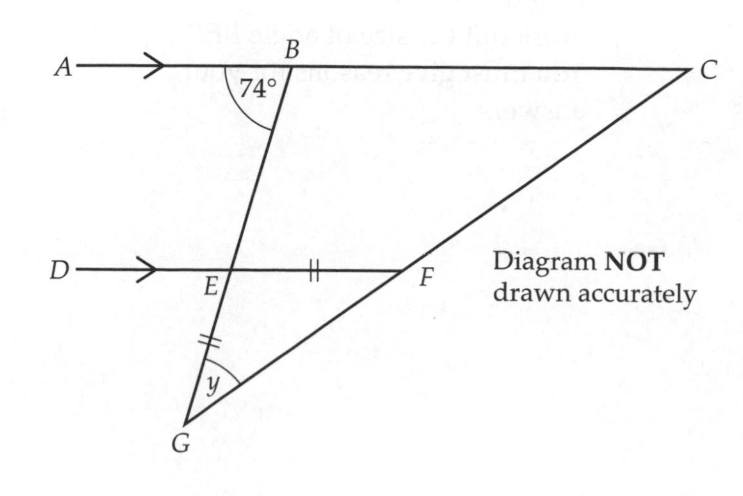

...................° **(4 marks)**

Angles in polygons

D 1 Calculate the size of an exterior angle of a regular decagon.

Exam questions similar to this have proved especially tricky – be prepared! ResultsPlus

Guided

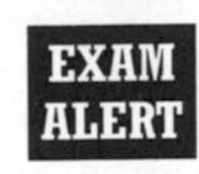

EXAM ALERT

Sum of exterior angles of any polygon = ...°

Exterior angle of a regular decagon $= \frac{.........}{10}$

= ....°

(2 marks)

D 2 A regular polygon has 15 sides.
Work out the size of an exterior angle of this regular polygon.

.....................° **(2 marks)**

D 3 The size of each exterior angle of a regular polygon is 45°.
Work out the number of sides of the polygon.

.... **(2 marks)**

D 4 The size of each interior angle of a regular polygon is 160°.
Work out the number of sides of the polygon.

Guided

Exterior angle of polygon = 180° − interior angle of polygon

= 180° − 160°

= ......°

Number of sides $= \frac{360}{.........}$

= ....

(2 marks)

D 5 A regular polygon has 24 sides.
Work out the size of each interior angle of this regular polygon.

.....................° **(2 marks)**

C 6 The diagram shows a regular pentagon.
Work out the size of the angle marked x.
Give reasons for your working.

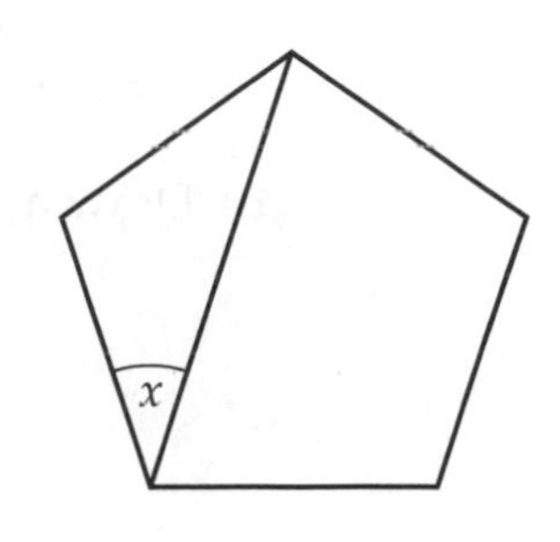

.....................° **(4 marks)**

Plan and elevation

D 1 Here are the plan, front elevation and side elevation of a 3-D shape.
Draw a sketch of the 3-D shape.

First copy the front elevation and draw in sloping lines for the sides.

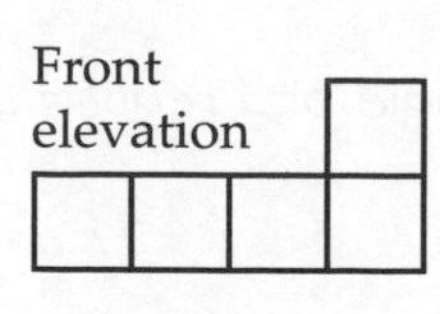

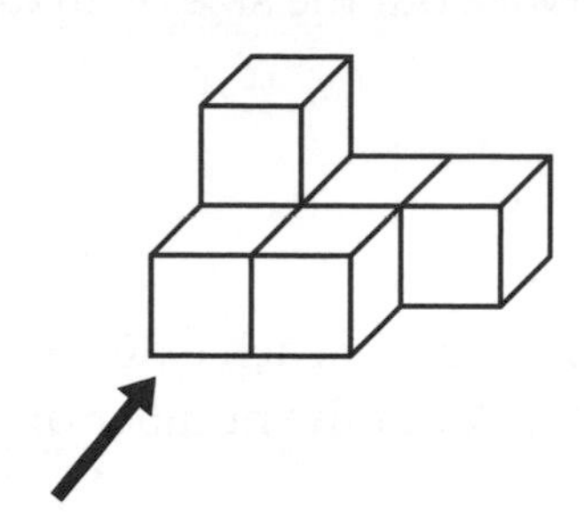

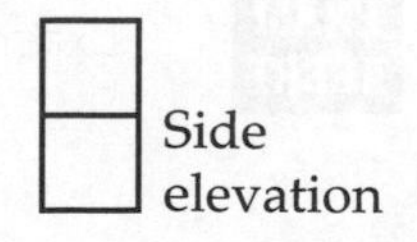

Guided

(2 marks)

D 2 The diagram shows a sketch of a solid object.
The solid object is made from 6 centimetre cubes.

(a) Draw a sketch of the elevation of the solid object in the direction marked by the arrow.

(2 marks)

(b) Draw a sketch of the plan of the solid object. **(2 marks)**

C 3 Here are the plan and front elevation of a solid shape.

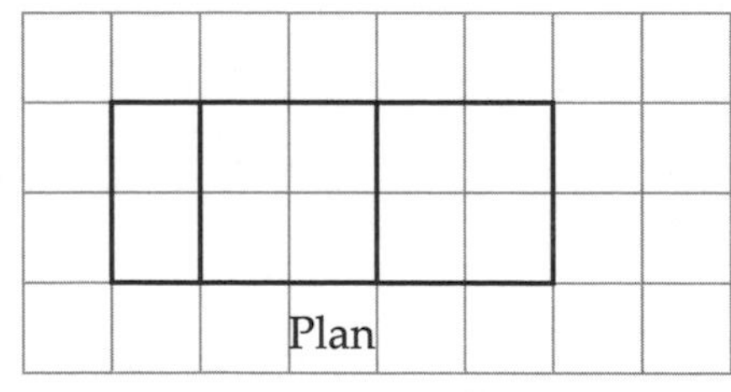

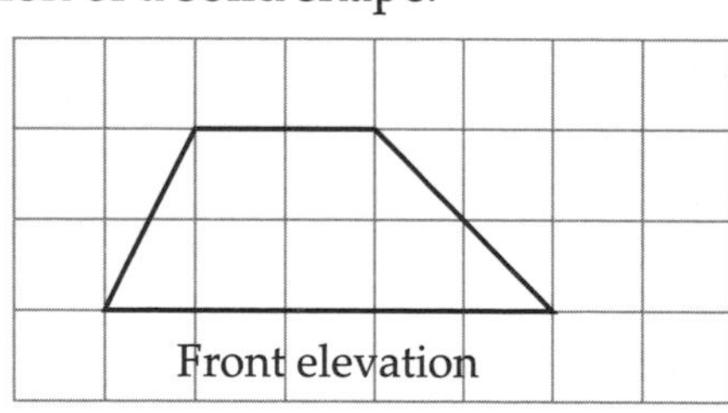

(a) On the grid below, draw the side elevation of the solid shape. **(2 marks)**

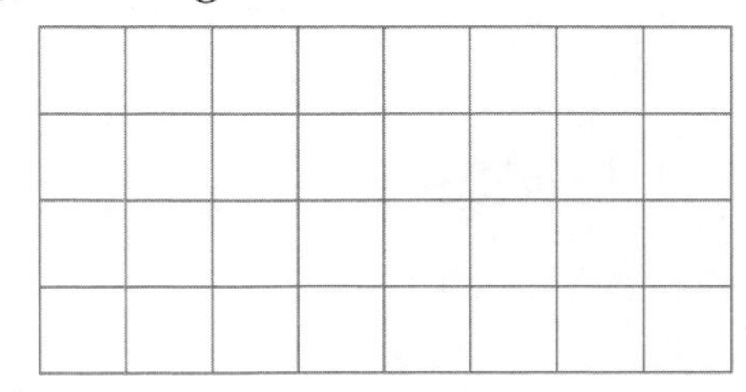

(b) Draw a 3-D sketch of the solid shape.

(2 marks)

Perimeter and area

D 1 Work out the area of this shape.

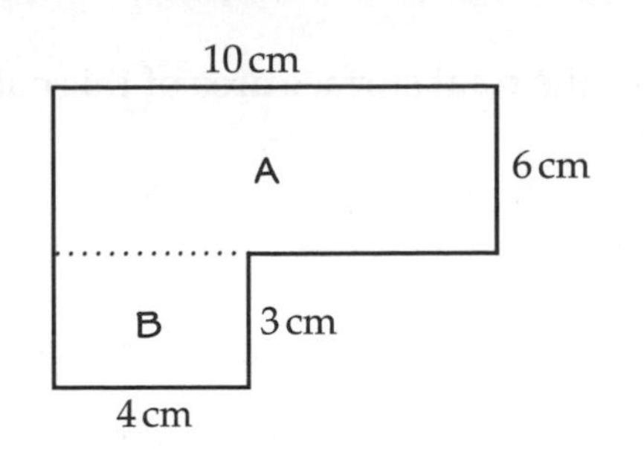

First divide the shape up into two rectangles.

Guided

Area of rectangle = length × width

Area of rectangle A = ... × ....

= .... cm²

Area of rectangle B = ... × ....

= .... cm²

Area of shape = ... + ... = ... cm²

(3 marks)

D 2 (a) Work out the area of this triangle.

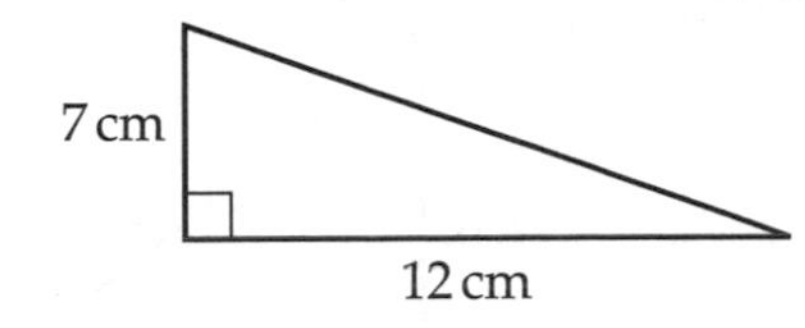

.... cm² **(2 marks)**

(b) Work out the area of this trapezium.

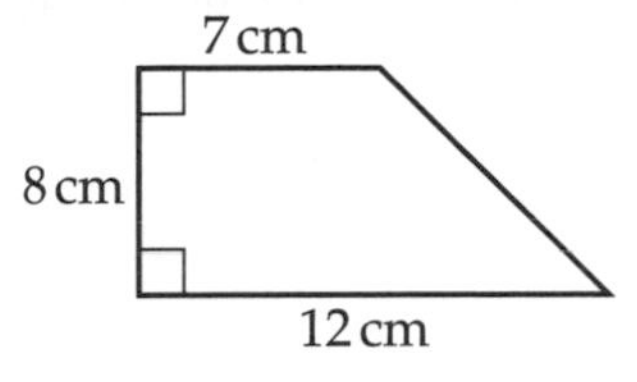

.... cm² **(3 marks)**

C 3 The diagram shows the plan of a room.
Mr Foster wants to cover the floor with varnish.
One tin of varnish will cover 20 m².
How many tins of varnish will he need?

First divide the shape up into a triangle and a rectangle.

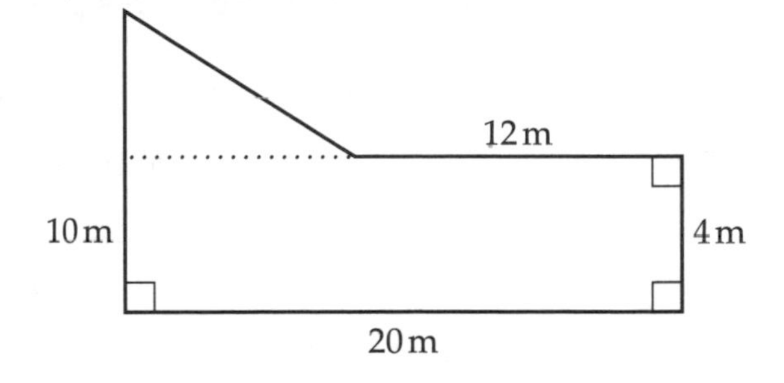

Guided

Area of rectangle = length × width

= .... × ....

= .... m²

Height of triangle = 10 − 4

= .... m

Area of triangle $= \frac{\text{base} \times \text{height}}{2}$

$= \frac{\text{.........} \times \text{.........}}{2}$ = ... m²

Area of shape = ... + ... = ... m²

Number of tins of varnish = ... ÷ 20 = ...tins

The answer must be a whole number of tins.

(5 marks)

C 4 The diagram shows the plan of a field.
The farmer wants to sell the field.
He wants to get at least £4 per square metre.
Mr Hobbs offers him £50 000 for the field.
Will the farmer accept his offer?

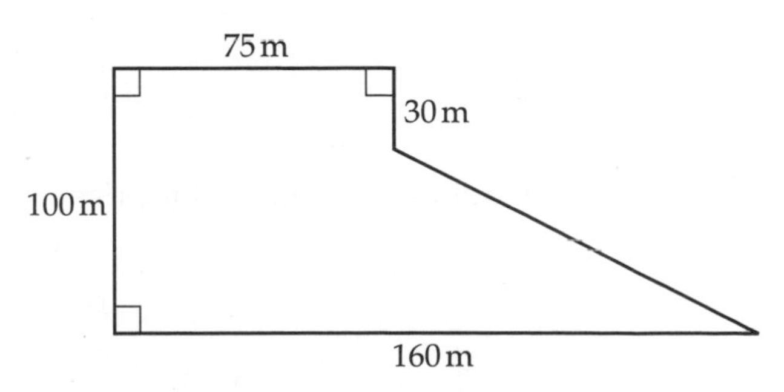

.... **(5 marks)**

Prisms

D 1 Work out the total surface area of this cuboid.

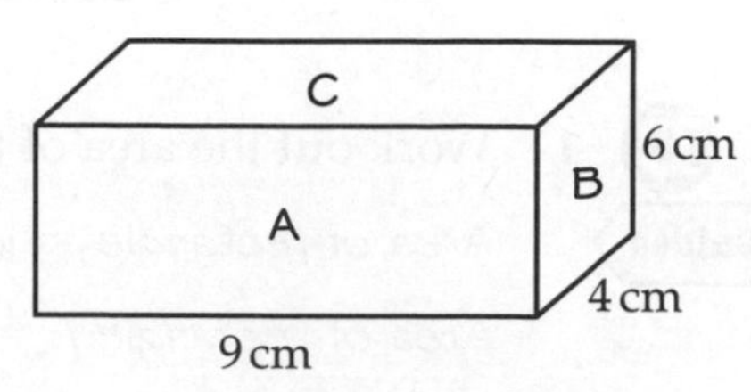

Guided Total surface area = 2 × area of face A + 2 × area of face B + 2 × area of face C

= 2 × .... × .... + 2 × .... × .... + 2 × .... × ....

= .... + ... + ...

= .... cm² **(2 marks)**

D 2 Work out the total surface area of this triangular prism.

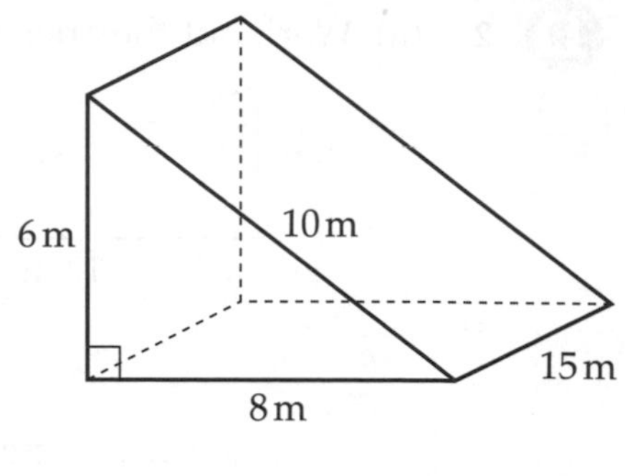

.... m² **(3 marks)**

C 3 The diagram shows a solid triangular prism.
Work out the volume of this prism.

Guided Volume of prism = area of cross-section × length

= area of triangle × length

= $\frac{1}{2}$ × .... × .... × ....

= .... cm³

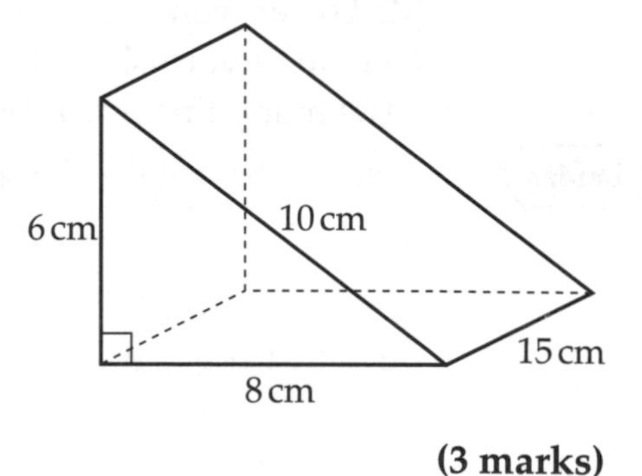

(3 marks)

C 4 The area of the cross-section of this prism is 16 cm².
The length of the prism is 20 cm.
Work out the volume of the prism.

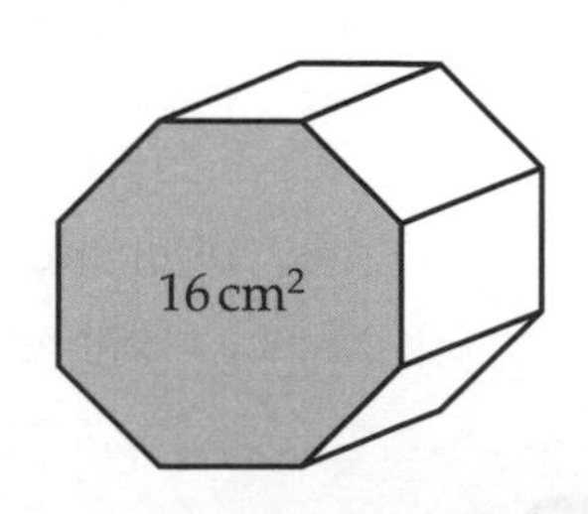

.... cm³ **(2 marks)**

C 5 Work out the volume of this prism.

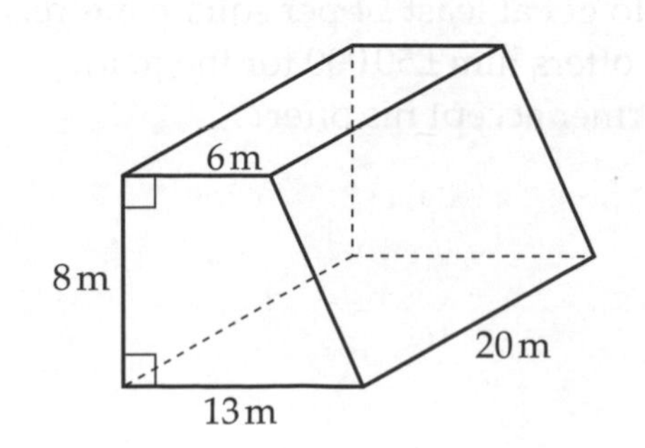

The cross-section is a trapezium.

.... m³ **(4 marks)**

Circles and cylinders

C **1** The diagram shows a semicircle.
The diameter of the semicircle is 18 cm.
Work out the perimeter of the semicircle.
Give your answer correct to 3 significant figures.

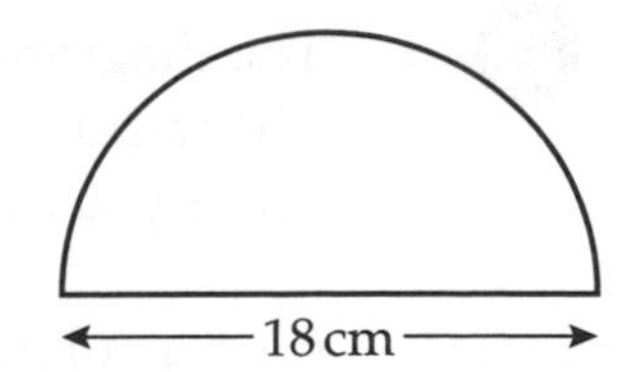

Perimeter of semicircle $= \frac{1}{2} \times$ circumference of circle + diameter

$= \frac{1}{2} \times \pi \times d + 18$

$= \frac{1}{2} \times \pi \times$ .... + 18

= ....

= .... cm correct to 3 s.f.

Exam questions similar to this have proved especially tricky – be prepared! ResultsPlus

(3 marks)

C **2** The diagram shows a cylinder of height 12 cm and radius 7 cm.

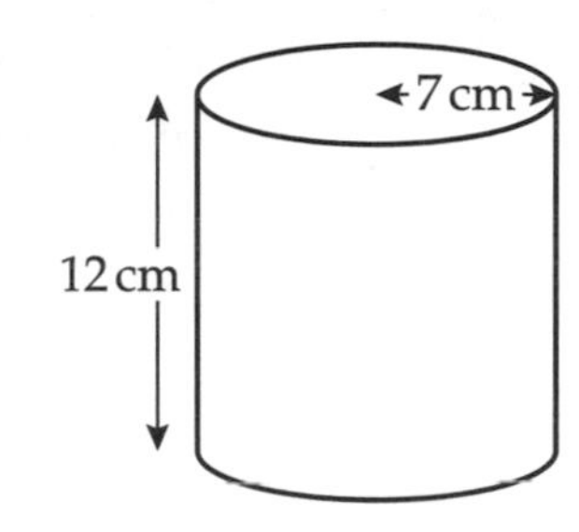

(a) Work out the volume of the cylinder.
Give your answer correct to 3 significant figures.

.... cm^3 **(2 marks)**

The cylinder is solid.

(b) Work out the **total** surface area of the cylinder.

The total surface area is made up of two circles and the curved surface area.

.... cm^2 **(3 marks)**

C **3** A circle has an area of 54 cm^2.
Work out the radius of the circle.
Give your answer correct to 3 significant figures.

Area $= \pi \times r^2$

.... $= \pi \times r^2$

$\frac{.........}{\pi} = r^2$

$r = \sqrt{.........}$

$r =$ ....

$r =$ .... cm correct to 3 s.f. **(2 marks)**

C **4** The diameter of a circle is 10 cm.
Work out the area of the circle.
Give your answer in terms of π.

.... **(2 marks)**

Sectors of circles

1 The diagram shows a sector of a circle, centre O.
The radius of the circle is 15 cm.
The angle of the sector is 140°.

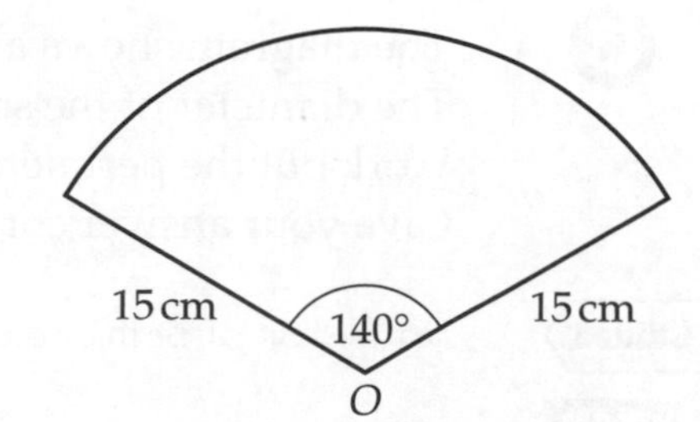

(a) Calculate the area of the sector.
Give your answer correct to 3 significant figures.

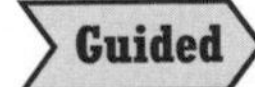

Area of sector $= \frac{\dots\dots\dots}{\dots\dots\dots} \times \pi \times \dots\dots\dots^2$

$= \dots\dots\dots\dots\dots\dots\dots$

$= \dots\dots\dots\dots\dots\dots\dots$ cm² correct to 3 s.f. **(2 marks)**

(b) Work out the perimeter of the sector.
Give your answer correct to 3 significant figures.

The perimeter consists of the arc length plus the length of two radii.

.... cm **(3 marks)**

A 2 The diagram shows a sector of a circle, centre O.
The radius of the circle is 5 cm.
The angle at the centre of the circle is 70°.
Find the perimeter of the sector.
Leave your answer in terms of π in its simplest form.

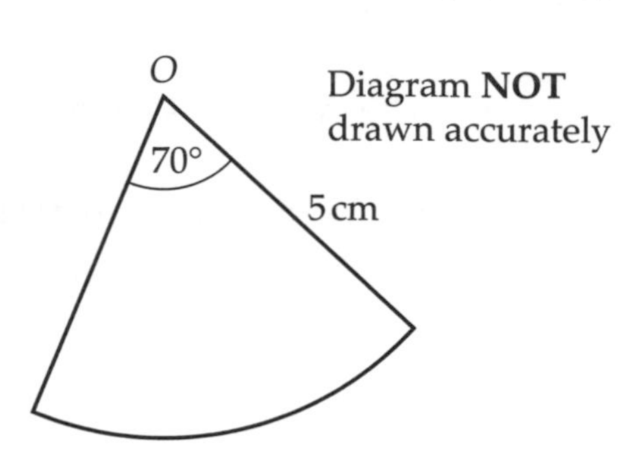

.... **(4 marks)**

A* 3 The diagram shows a sector of a circle, centre O, radius 8 cm.
The arc length of the sector is 7 cm.
Calculate the area of the sector.

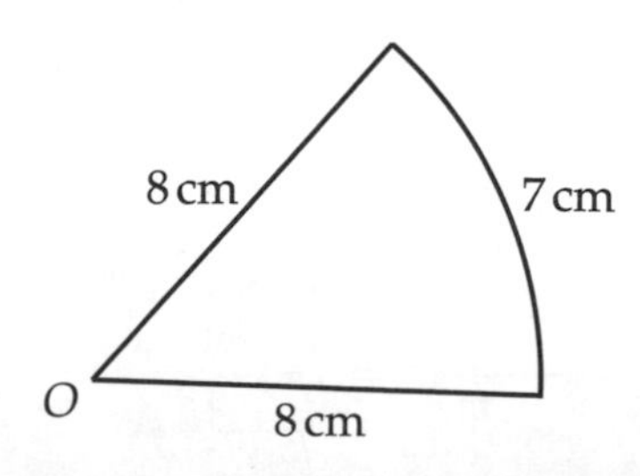

Use the arc length to find the angle of the sector.
Then use your angle to work out the area of the sector.

.... cm² **(4 marks)**

Volumes of 3-D shapes

Remember to use the formula sheet to look up volume formulae.

A **1** A sphere has a radius of 6 cm.
Work out the volume of the sphere.
Give your answer as a multiple of π.

Volume of sphere $= \frac{4}{3}\pi \times r^3$

$= \frac{4}{3} \times \pi \times \text{.........}^3$

$= \text{.....................}\pi\text{ cm}^3$ **(2 marks)**

A **2** A cone has a base radius of 8.5 cm and a height of 24 cm.
Work out the volume of the cone.
Give your answer correct to 3 significant figures.

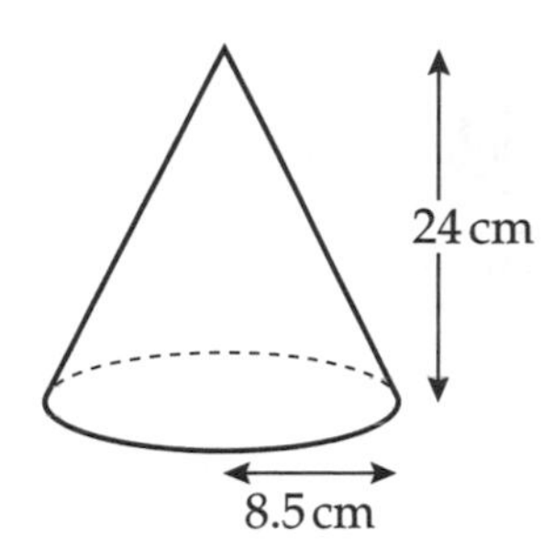

.... cm^3 **(2 marks)**

A* **3** A cone has 5 cm cut off its top as shown.
The shape which is left is known as a frustum.
Calculate the volume of the frustum to 3 significant figures.

Guided

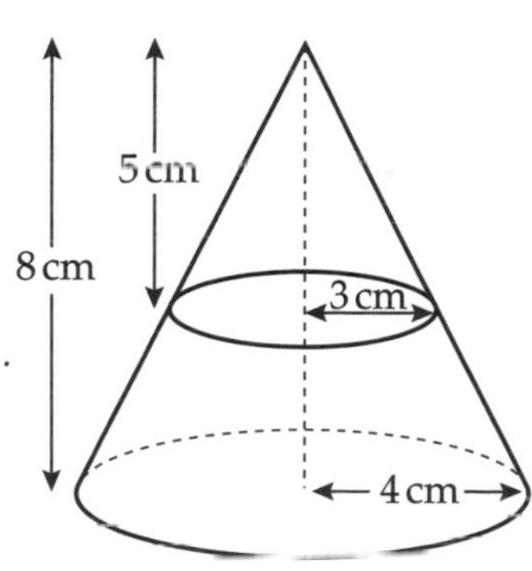

Volume of original cone $= \frac{1}{3}\pi r^2 h = \frac{1}{3} \times \pi \times 4^2 \times 8 =$ ......

Volume of 'missing' cone $= \frac{1}{3}\pi r^2 h = \frac{1}{3} \times \pi \times \text{......}^2 \times \text{......} = \text{......}$

Volume of frustum

= volume of original cone − volume of 'missing' cone

= .... − ....

= .... cm^3 correct to 3 s.f. **(4 marks)**

A* **4** A cone is joined to a hemisphere as shown in the diagram.
Work out the volume of the shape.
Give your answer as a multiple of π.

A hemisphere is half a sphere.

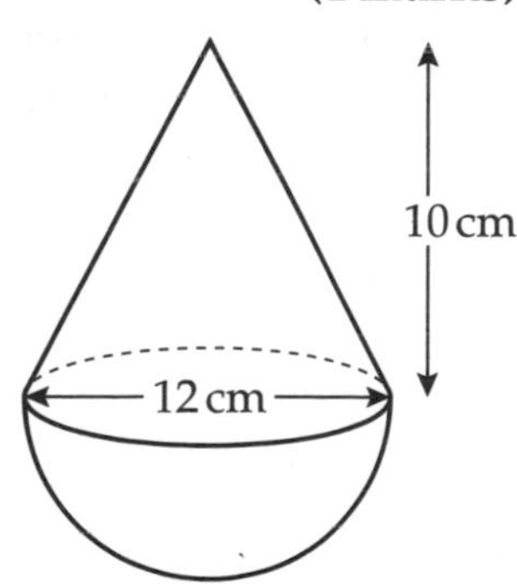

.... cm^3 **(4 marks)**

 **5** A clay bowl is in the shape of a hollow hemisphere.
The external radius of the bowl is 10.4 cm.
The internal radius of the bowl is 9.8 cm.
Work out the volume of clay in the bowl.
Give your answer correct to 3 significant figures.

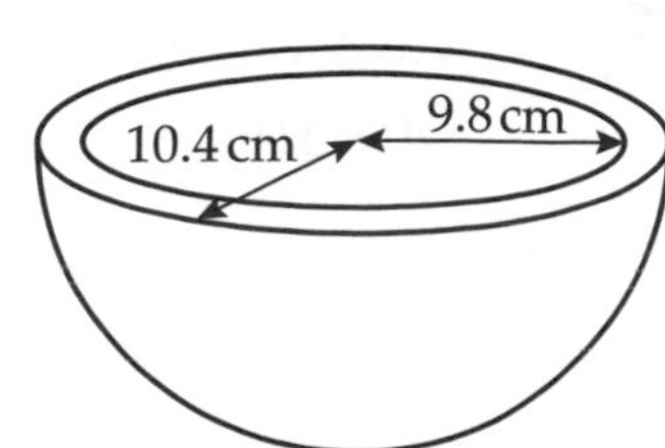

.... cm^3 **(3 marks)**

Pythagoras' theorem

C 1 Work out the length of PQ.
Give your answer correct to 3 significant figures.

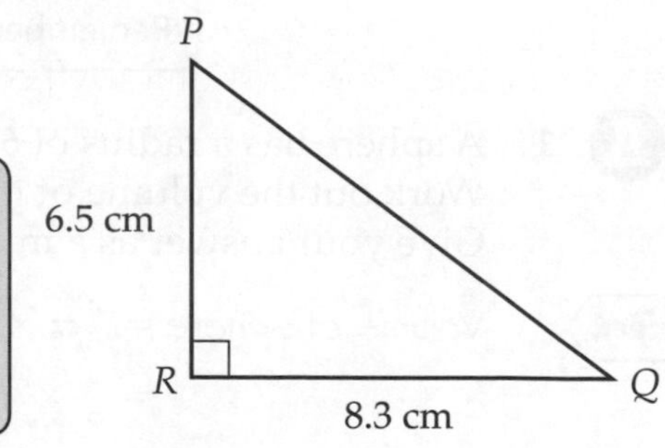

Guided

$PQ^2 = 6.5^2 +$ ....

$PQ^2 =$ ....

$PQ = \sqrt{\ldots\ldots\ldots}$

$PQ =$ ....

$PQ =$ .... cm correct to 3 s.f. **(3 marks)**

Pythagoras' theorem

$a^2 + b^2 = c^2$

C 2 Work out the length of DE.

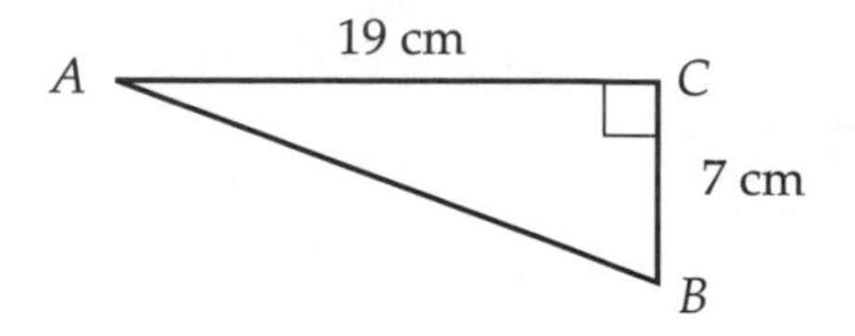

Guided

$DE^2 +$ .... $= 26^2$

$DE^2 = 26^2 -$ ....

$DE^2 =$ ....

$DE = \sqrt{\ldots\ldots\ldots}$

$DE =$ .... cm **(3 marks)**

The hypotenuse is 26 cm so you are finding one of the shorter sides.

C 3 Work out the length of AB.
Give your answer correct to 3 significant figures.

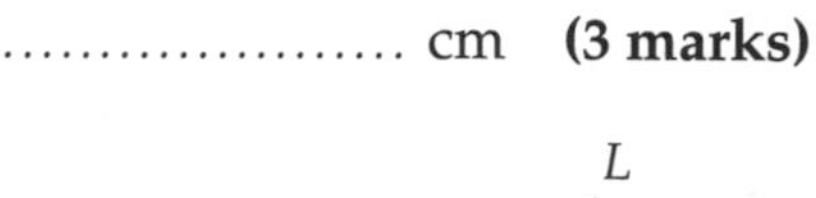

.... cm **(3 marks)**

C 4 Work out the length of LM.
Give your answer correct to 3 significant figures.

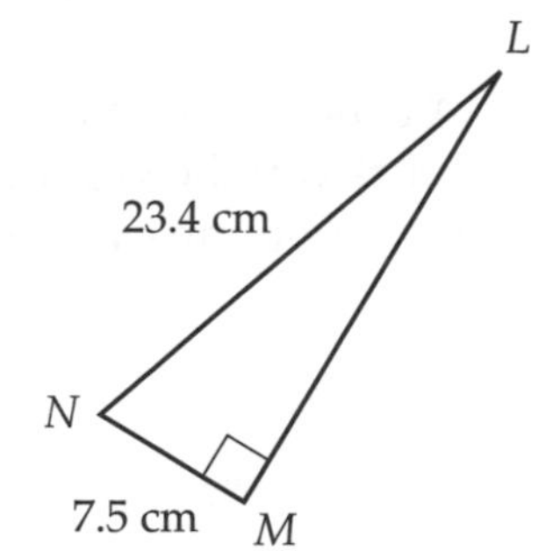

.... cm **(3 marks)**

C 5 Work out the height of the isosceles triangle.
Give your answer correct to 1 decimal place.

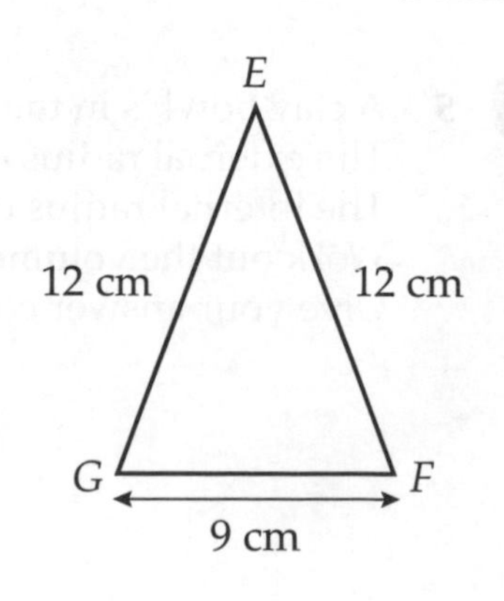

.... cm **(3 marks)**

Surface area

A **1** A sphere has a radius of 10 cm.
Work out the total surface area of the sphere.
Give your answer correct to 3 significant figures.

Guided

Surface area of a sphere $= 4\pi r^2$

$= 4 \times \pi \times \text{.........}^2$

$=$ ....

$=$ .... cm^2 correct to 3 s.f. **(2 marks)**

A **2** A solid shape is made from a cone on top of a hemisphere.
The base of the cone has a radius of 6 cm and a vertical height of 8 cm.
The hemisphere has a radius of 6 cm.
Work out the surface area of the shape.
Give your answer correct to 3 significant figures.

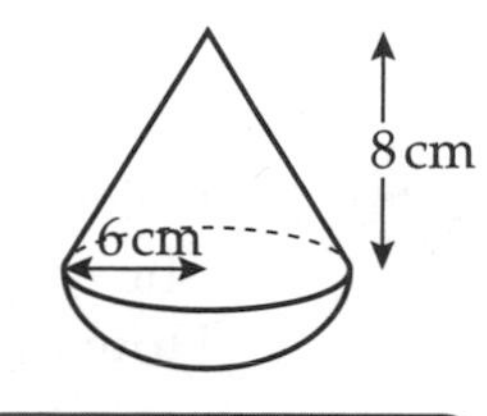

Use Pythagoras' theorem to work out the slant height of the cone, l.

.... cm^2 **(5 marks)**

A* **3** The radius of a sphere is 4 cm.
The radius of the base of a cone is 6 cm.
The surface area of the sphere is equal to the curved surface area of the cone.
Work out the slant height of the cone.

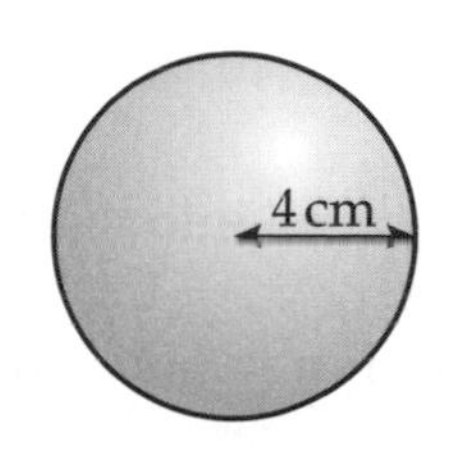

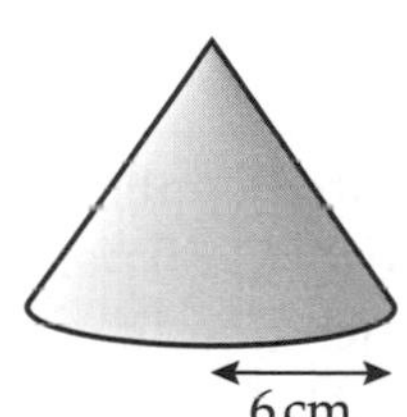

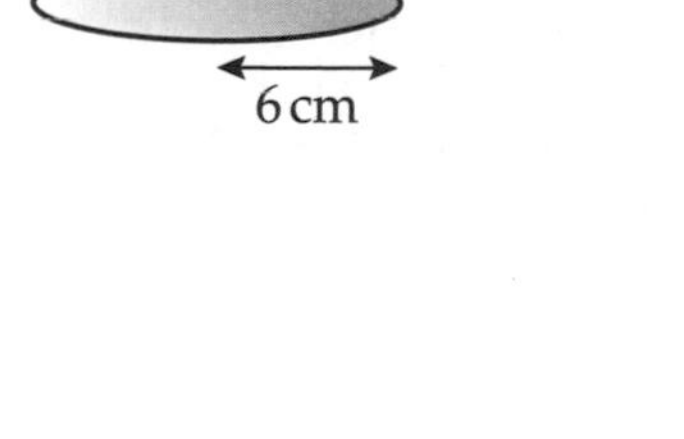

Guided

Surface area of sphere = curved surface area of cone

$4\pi r^2 = \pi r l$

$4 \times \pi \times \text{.........}^2 = \pi \times \text{.........} \times l$

$\dfrac{\text{............}}{\text{............}} = l$

$l = \dfrac{\text{.........}}{\text{.........}}$ cm **(4 marks)**

A* **4** The radius of a sphere is 3 cm.
The radius of the base of a cone is also 3 cm.
The surface area of the sphere is equal to the total surface area of the cone.
Work out the slant height of the cone.

$l =$ .... cm **(4 marks)**

Converting units

C **1** A turkey weighs 5 kg.
Mark knows that he must cook the turkey for 20 minutes for every pound.
For how long should Mark cook the turkey?
Give your answer in hours and minutes.

Guided Weight of turkey in pounds = 5 × 2.2

= .... pounds

Cooking time = ... × 20

= .... minutes

= .... hours and .... minutes **(3 marks)**

C **2** Helen needs to buy a 5.4 metre length of carpet.
The carpet costs £12.95 per foot.
How much will the carpet cost?

You should know that 30 cm ≈ 1 foot.

£..................... **(3 marks)**

C **3** Bananas in shop A cost £1.65 for 1.5 kg.
Bananas in shop B cost £0.55 per pound.
In which shop are the bananas better value for money?

.... **(4 marks)**

C **4** Kalim puts 10 gallons of petrol into his car's petrol tank.
Petrol costs £1.30 per litre.
How much does it cost Kalim to fill up his petrol tank?

You should know the fact that 1 gallon = 4.5 litres.

Guided 10 gallons = 10 × 4.5 litres

= .... litres

Cost = £1.30 × ....

= £......... **(3 marks)**

C **5** Sue drives at an average speed of 80 km/hour.
How long will it take her to drive 100 miles?

.... hours **(3 marks)**

C **6** Lottie buys a 1.5 litre bottle of drink.
She pours all the drink into glasses.
Each glass has a capacity of $\frac{1}{2}$ pint.
How many glasses will she be able to fill?

.... **(3 marks)**

Units of area and volume

C **1** Change 7 cm^3 into mm^3.

Guided

EXAM ALERT

$1\ cm = 10\ mm$

$1\ cm^3 = 10^3\ mm^3$

$= \dots\dots\ mm^3$

$7\ cm^3 = 7 \times \dots\dots\ mm^3$

$= \dots\dots\ mm^3$

(2 marks)

Exam questions similar to this have proved especially tricky – be prepared! ResultsPlus

C **2** Change 7.5 m^2 into cm^2.

.... cm^2 **(2 marks)**

C **3** Change 4 200 000 cm^2 into m^2.

You are changing a smaller unit to a larger unit so divide by a power of 10.

.... m^2 **(2 marks)**

B **4** A tank has a volume of 6 m^3.
Work out how many litres of water the tank contains.

Guided

$1\ m = 100\ cm$

$1\ m^3 = \dots\dots\dots\ cm^3$

$6\ m^3 = 6 \times \dots\dots\dots$

$= \dots\dots\dots\ cm^3$

Now use the fact that 1 litre = 1000 cm^3 to work out how many litres of water the tank holds.

.... litres **(3 marks)**

B **5** A water trough is in the shape of a cuboid.
The trough contains 600 litres of water when full.
The trough has a length of 1.5 m and a width of 80 cm.
Work out the height of the trough.

.... cm **(4 marks)**

B **6** A large container in the shape of a cuboid has a volume of 0.384 m^3.
The container is filled with water from cartons.
Each carton is in the shape of a cube of side 40 cm.
How many cartons of water will fit in the container?

.... **(3 marks)**

Speed

D **1** The distance from London to Munich is 900 km.
A flight from London to Munich takes 3 hours.
Work out the average speed of the aeroplane.

Guided Speed = $\frac{\text{distance}}{\text{time}}$

= $\frac{\dots\dots\dots}{\dots\dots\dots}$

= .... km/h **(2 marks)**

C **2** Lottie drives at an average speed of 80 km/h.
Her journey takes $2\frac{1}{2}$ hours.
Work out the distance Lottie drives.

.... km **(2 marks)**

C **3** A train travelled 180 km in $1\frac{1}{2}$ hours.
Work out the average speed of the train.

.... km/h **(2 marks)**

C **4** An aeroplane travelled at an average speed of 600 km/h for 2 hours 15 minutes.
How far did the aeroplane travel?

Use the fact that 15 minutes is $\frac{1}{4}$ of an hour.

.... km **(2 marks)**

C **5** Jamil walked at an average speed of 6 km/h for 40 minutes.
How far did he walk?

Guided Time = $\frac{40}{60}$ hour

Write the time as a fraction of an hour and simplify the fraction.

= $\frac{\dots\dots\dots}{\dots\dots\dots}$ hour

Distance = speed × time

= 6 × $\frac{\dots\dots\dots}{\dots\dots\dots}$

= .... km **(2 marks)**

C **6** A car travels at an average speed of 90 km/h.
How far will the car travel in 20 minutes?

.... km **(2 marks)**

C **7** Jim leaves home at 14:30
He cycles 39 km to a friend's house at an average speed of 18 km/h.
At what time does he arrive at his friend's house?

.... **(2 marks)**

Density

C **1** (a) The mass of $3\,m^3$ of zinc is 21 390 kg.
Work out the density of zinc.

Guided

Density = $\frac{\text{mass}}{\text{volume}}$

Density = $\frac{\dots\dots\dots}{\dots\dots\dots}$

= .... kg/m^3 **(2 marks)**

(b) The density of copper is $8960\,kg/m^3$.
Work out the mass of $1.5\,m^3$ of copper.

Guided

Mass = density × volume

Mass = ... × ....

= .... kg **(2 marks)**

C **2** The diagram shows a solid cuboid.
The cuboid has length 8 cm, width 6 cm and height 5 cm.
The cuboid is made of wood.
The wood has a density of 0.54 grams per cm^3.
Work out the mass of the cuboid.

First find the volume of the cuboid.

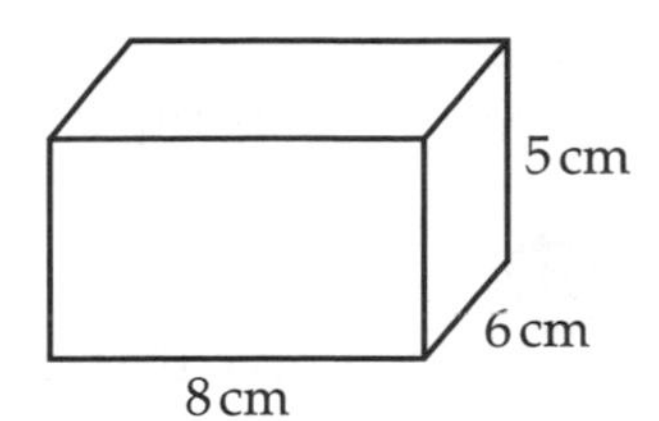

.... grams **(3 marks)**

B **3** The density of gold is 19.3 grams per cm^3.
A gold ingot is in the shape of a cuboid.
The cuboid has length 20 cm, width 8 cm and height 1.5 cm.
Work out the mass of the gold ingot.
Give your answer in kilograms.

.... kg **(3 marks)**

B **4** A solid cylinder is made of wood.
It has a radius of 8 cm and a height of 3 cm.
The cylinder has a mass of 392 grams.
Work out the density of the wood.
Give your answer correct to 2 significant figures.

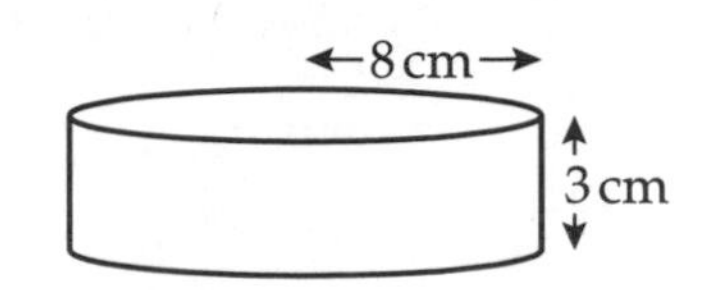

.... g/cm^3 **(3 marks)**

Congruent triangles

1 *ABCD* is a quadrilateral.
AB is parallel to *DC*.
DA is parallel to *CB*.
Prove that triangle *ABC* is congruent to triangle *CDA*.

First draw in line *AC* to create the two triangles *ABC* and *CDA*.

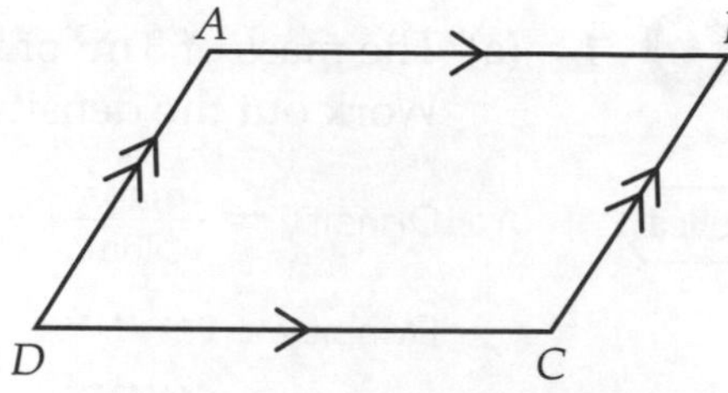

EXAM ALERT

Statement	Reason
Angle *CAB* = Angle *ACD*	....
Angle *BCA* = Angle *DAC*	....
AC = *AC*	....

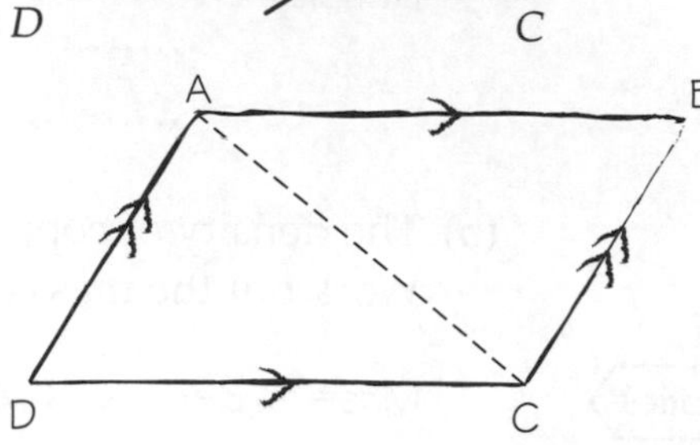

There is more than one way to prove that these two triangles are congruent.

Exam questions similar to this have proved especially tricky – be prepared! ResultsPlus

So triangle .... is congruent to triangle .... (AAS) **(3 marks)**

2 *ABCDE* is a regular pentagon.
Prove that triangle *ADE* is congruent to triangle *CED*.

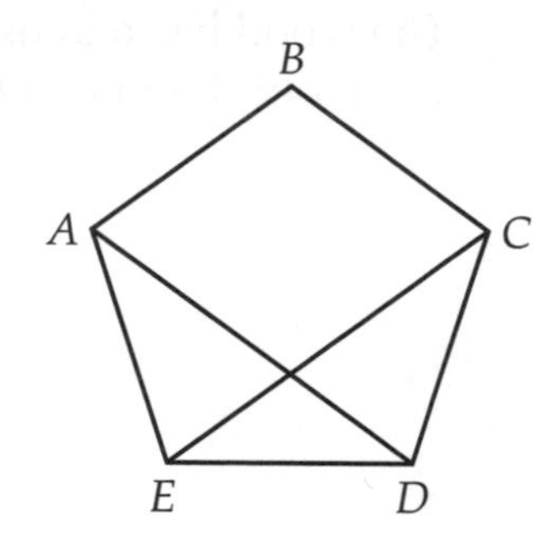

(3 marks)

3 *ABC* is an isosceles triangle.
D lies on *BC*.
AD is perpendicular to *BC*.
Prove that triangle *ADC* is congruent to triangle *ADB*.

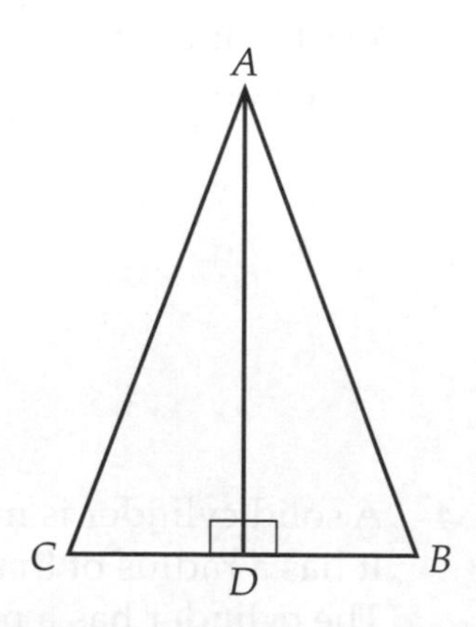

(3 marks)

Similar shapes 1

1 The two triangles *ABC* and *PQR* are mathematically similar.

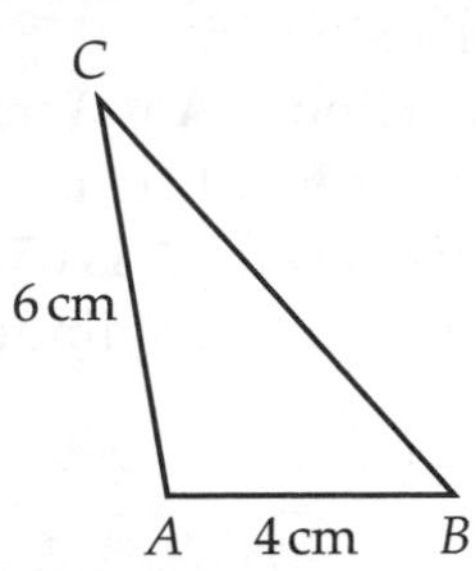

R
10 cm
P
6 cm
Q

(a) Work out the length of *PR*.

Guided

$\frac{PR}{AC} = \frac{PQ}{AB}$ so $\frac{PR}{\ldots\ldots} = \frac{\ldots\ldots}{\ldots\ldots}$

$PR = \frac{\ldots\ldots}{\ldots\ldots} \times \ldots\ldots$

$= \ldots\ldots\ldots\ldots$ cm

Fill in the lengths of the sides you know.

You can also use scale factors for problems of this type.

(2 marks)

(b) Work out the length of *BC*.

Guided

$\frac{BC}{QR} = \frac{AB}{PQ}$ so $\frac{BC}{\ldots\ldots} = \frac{\ldots\ldots}{\ldots\ldots}$

$BC = \frac{\ldots\ldots}{\ldots\ldots} \times \ldots\ldots$

$= \ldots\ldots\ldots\ldots$ cm

(2 marks)

A

2 *BE* is parallel to *CD*.

Work out the length of *CD*.

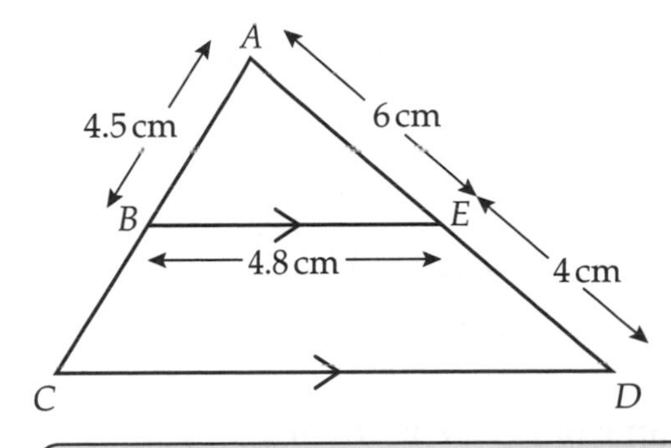

Diagram **NOT** drawn accurately

The triangles *AEB* and *ADC* are similar. Start by sketching these triangles and adding the dimensions.

.... cm **(2 marks)**

3 *AE* is parallel to *CD*.

Work out the length of *AD*.

Triangle *ABE* is similar to triangle *DBC*. Start by sketching the triangles side by side.

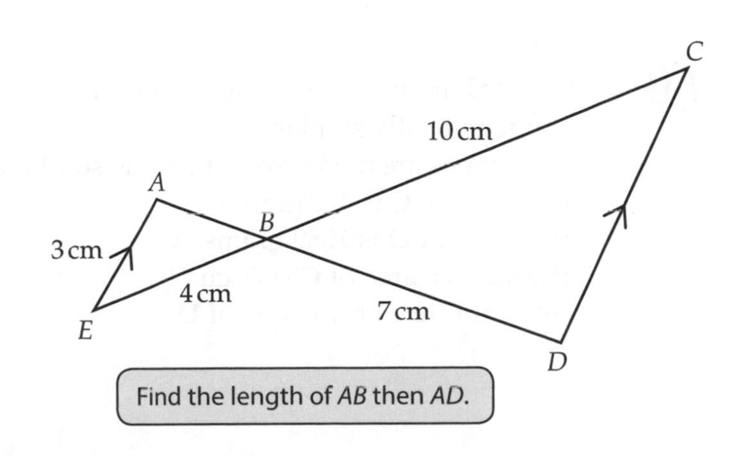

Find the length of *AB* then *AD*.

.... cm **(3 marks)**

Similar shapes 2

1 Cylinder **A** and cylinder **B** are mathematically similar.
The length of cylinder **A** is 3 cm and the length of cylinder **B** is 12 cm.
The volume of cylinder **A** is 65 cm^3.
Calculate the volume of cylinder **B**.

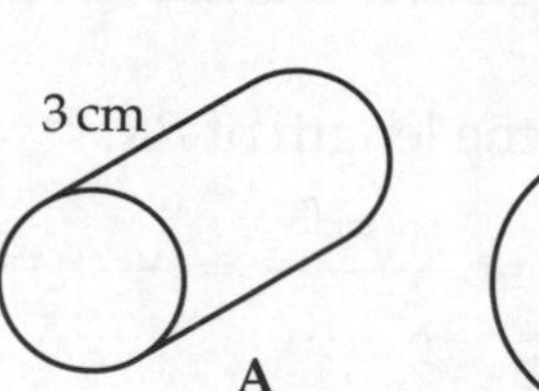

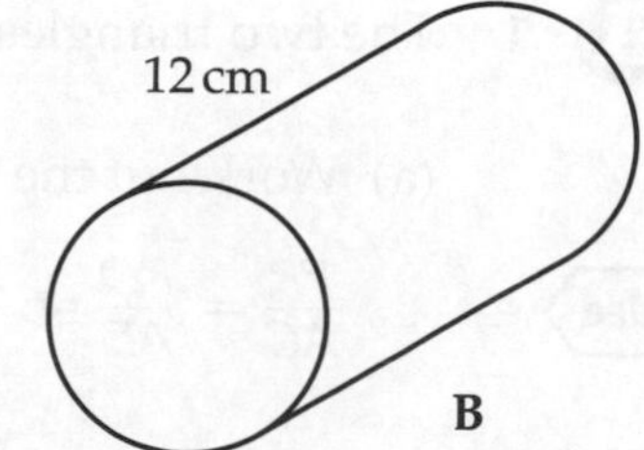

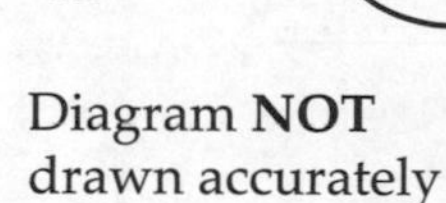

Diagram **NOT** drawn accurately

Guided

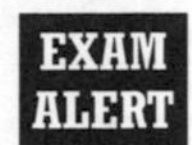

Linear scale factor $= \frac{12}{\dots\dots\dots} = \dots\dots\dots$

Volume scale factor $= \dots\dots\dots^3 = \dots\dots\dots$

Volume of cylinder **B** = ... × .... = ... cm^3

(3 marks)

Exam questions similar to this have proved especially tricky – be prepared! ResultsPlus

2 Two cones, **P** and **Q**, are mathematically similar.
The total surface area of cone **P** is 28 cm^2.
The total surface area of cone **Q** is 112 cm^2.
The height of cone **P** is 5 cm.

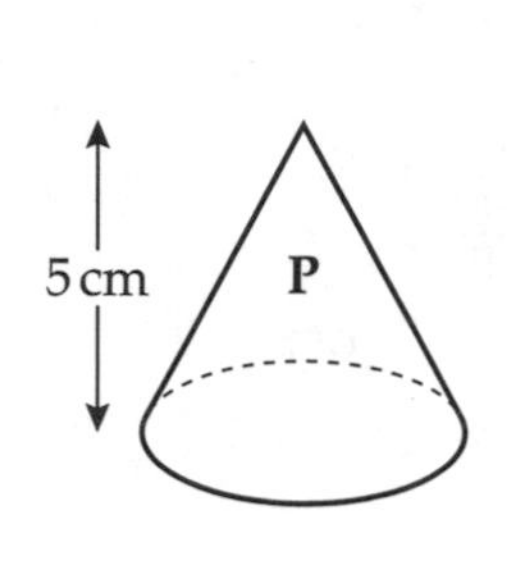

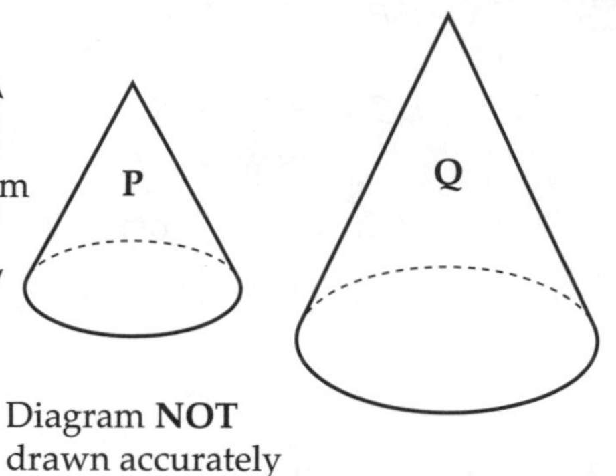

Diagram **NOT** drawn accurately

(a) Work out the height of cone **Q**.

Guided

Area scale factor $= \frac{112}{\dots\dots\dots} = \dots\dots\dots$

Length scale factor $= \sqrt{\dots\dots\dots} = \dots\dots\dots$

Height of cone **Q** = ... × 5 = ... cm

(3 marks)

The volume of cone **P** is 16 cm^3.

(b) Work out the volume of cone **Q**.

Guided

Volume scale factor = (linear scale factor)^3

= (.........)^3 = ...

Volume of **Q** = ... × .... = ... cm^3

(2 marks)

3 **C** and **D** are two solid shapes which are mathematically similar.
The shapes are made from the same solid material.
The mass of **C** is 320 grams.
The mass of **D** is 1080 grams.
The surface area of **C** is 76 cm^2.
Work out the surface area of **D**.

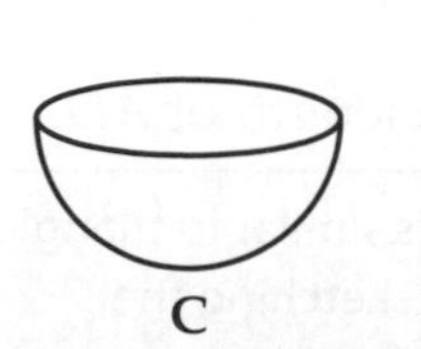

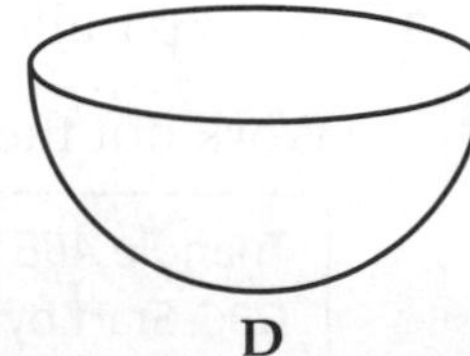

Use the masses to find the volume scale factor.

.... cm^2 **(4 marks)**

Bearings

D 1 Measure and write down the bearing of B from A.

Guided

> Bearings are measured clockwise from North.
> Start by marking the angle you need to find.
> Measure the acute angle then subtract it from 360°.

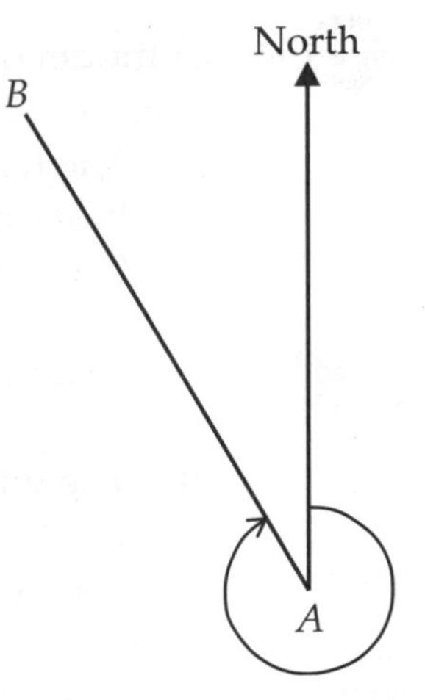

....................° **(1 mark)**

C 2 The bearing of Q from P is 140°.
What is the bearing of P from Q?

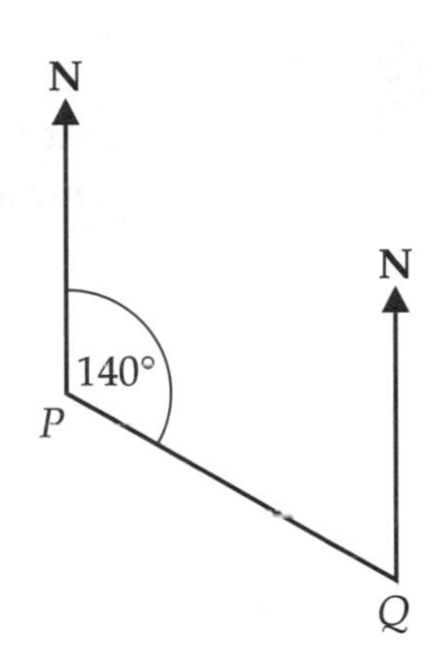

....................° **(2 marks)**

C 3 The diagram shows the positions of two boats, P and Q.

The bearing of a boat R from boat P is 030°.
The bearing of boat R from boat Q is 300°.
In the space above, draw an accurate diagram to show the position of boat R.
Mark the position of boat R with a cross (×). Label it R.

(3 marks)

Scale drawings and maps

D 1 A model of an aeroplane is made using a scale of 1 : 500

(a) The length of the model is 14.2 cm.
Work out the length of the real aeroplane.
Give your answer in metres.

Guided Length of aeroplane = 14.2 × .... = ... cm = ... m **(2 marks)**

(b) The width of the real aeroplane is 59.5 m.
Work out the width of the model.
Give your answer in centimetres.

Guided Width of model = 59.5 ÷ .... = ... m = ... cm **(2 marks)**

C 2 The diagram shows two towns A and B.
Work out the distance from town A to town B.
Give your answer in kilometres.

Scale: 1 cm represents 20 km

× A

× B

.... km **(2 marks)**

C 3 The diagram shows the positions of two boats, P and Q.

N

N

× Q

× P

Scale 1 cm represents 5 km

(a) How far away from boat P is boat Q?
Give your answer in kilometres.

.... km **(2 marks)**

(b) The bearing of a boat R from boat P is 070°. The bearing of boat R from boat Q is 300°.
How far away from boat P is boat R?
Give your answers in kilometres.

.... km **(3 marks)**

Constructions

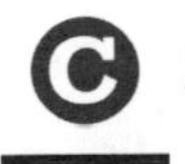

1 Use ruler and compasses to **construct** the perpendicular to the line segment AB that passes through the point P.
You must show all your construction lines.

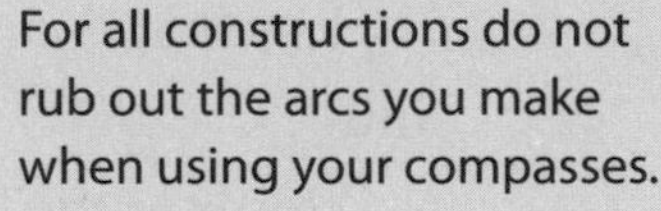

(2 marks)

2 Use ruler and compasses to construct the perpendicular bisector of the line AB.
You must show all your construction lines.

Exam questions similar to this have proved especially tricky – be prepared!

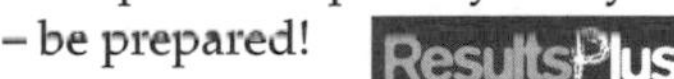

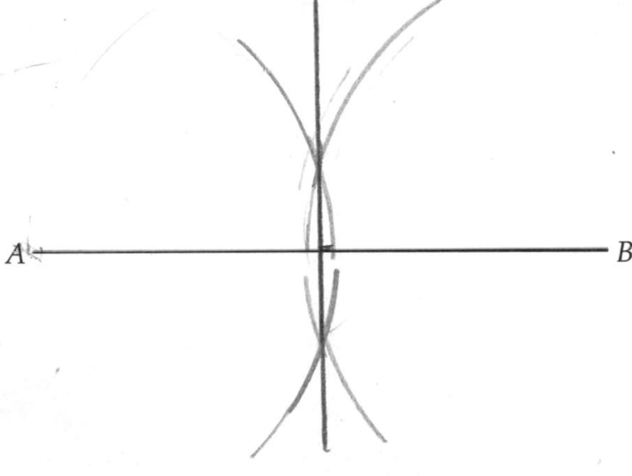

(2 marks)

3 Use ruler and compasses to construct the bisector of angle RPQ.
You must show all your construction lines.

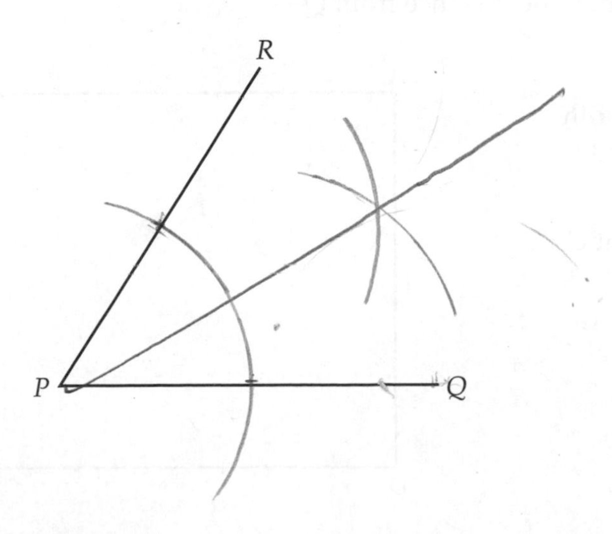

(2 marks)

Loci

 1 *ABC* is a triangle.
Shade the region inside the triangle which is **both**
less than 5 cm from the point *B*
and
closer to the line *AB* than the line *AC*.

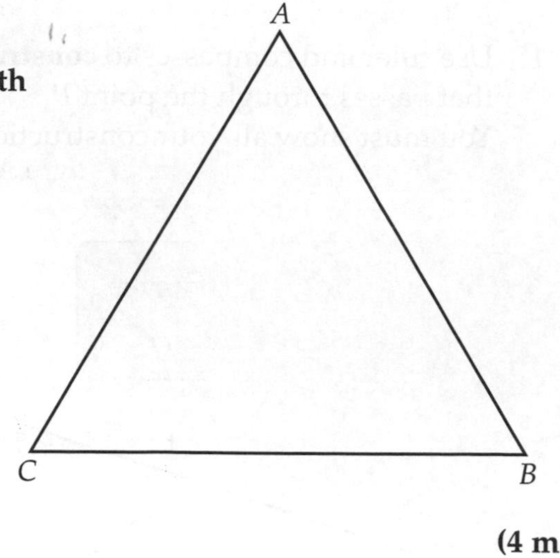

The question does **not** use the word 'construct' so you can bisect angle *CAB* by using a protractor to measure the angle.

(4 marks)

2 The diagram shows two points, *P* and *Q*.

P× ×*Q*

On the diagram shade the region that contains all the points that satisfy **both** the following:
the distance from *P* is less than 4 cm
the distance from *P* is greater than the distance from *Q*

(4 marks)

 3 *ABCD* is a rectangle.
Shade the set of points which are **both**
more than 2 cm from the line *AB*
and
more than 3.5 cm from the point *C*.

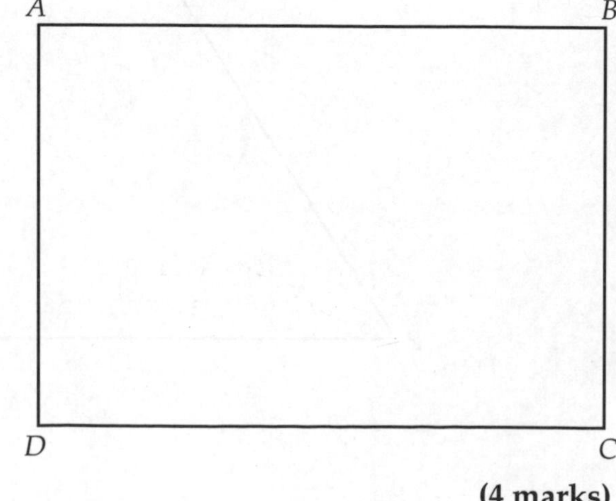

(4 marks)

Translations, reflections and rotations

1 (a) On the grid, translate triangle **A** by $\begin{pmatrix} 4 \\ -3 \end{pmatrix}$.
Label the new triangle **B**.

$\begin{pmatrix} 4 \\ -3 \end{pmatrix}$ means 4 units to the right and 3 units down.

(2 marks)

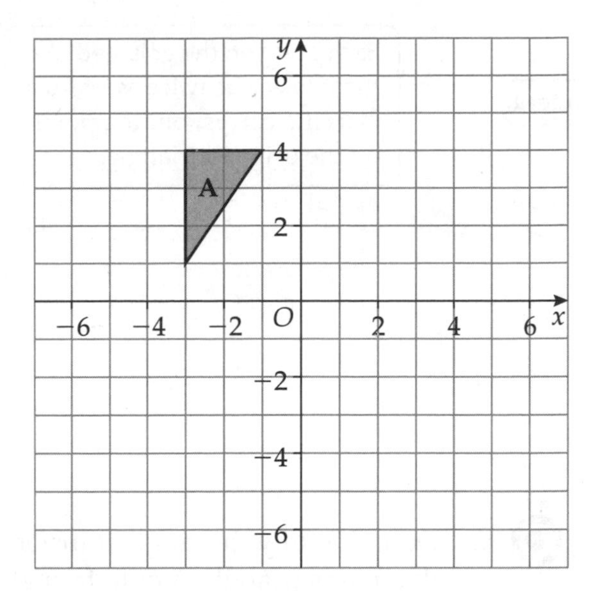

(b) On the grid, rotate triangle **A** 180° about (0, 0).
Label the new triangle **C**.

Use tracing paper to help with the rotation.

(2 marks)

2 (a) Describe fully the single transformation that will map triangle **A** onto triangle **B**.

EXAM ALERT

Exam questions similar to this have proved especially tricky – be prepared! ResultsPlus

....

....

(2 marks)

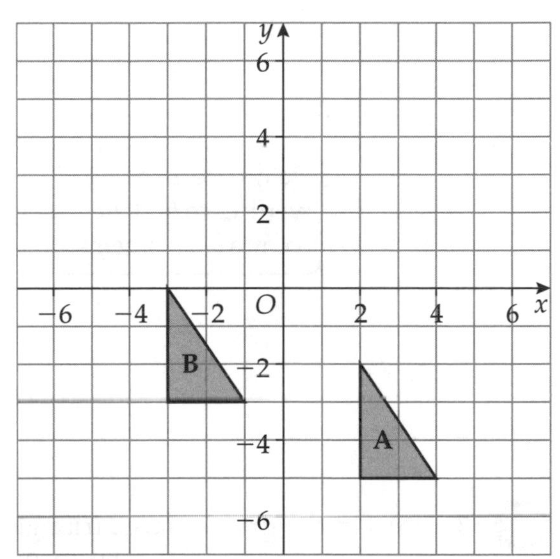

(b) On the grid, rotate triangle **A** 180° about (1, 0).
Label the new triangle **C**.

(2 marks)

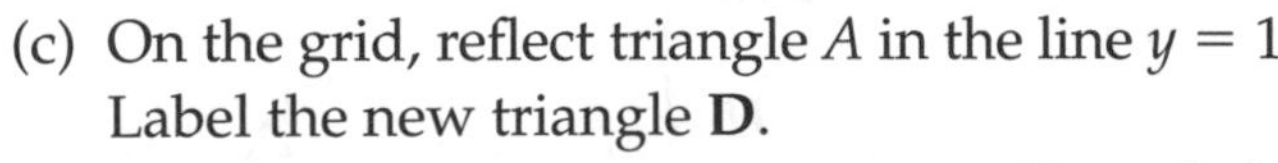

(c) On the grid, reflect triangle *A* in the line $y = 1$
Label the new triangle **D**.

(2 marks)

3 (a) Describe fully the single transformation that will map triangle **P** onto triangle **Q**.

....

....

(2 marks)

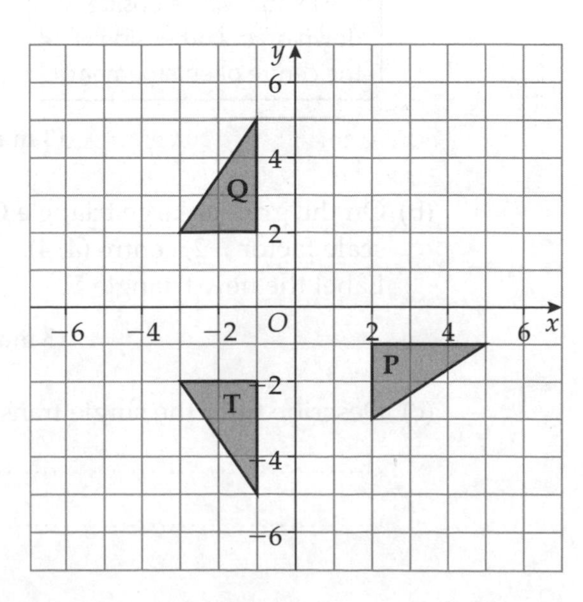

(b) Describe fully the single transformation that will map triangle **P** onto triangle **T**.

....

....

(3 marks)

Enlargements

C **1** On the grid, enlarge the shape by a scale factor of 2, centre (0, 0).

Guided

Each point on the enlarged triangle will be twice as far from O as the corresponding point on the original triangle.

(3 marks)

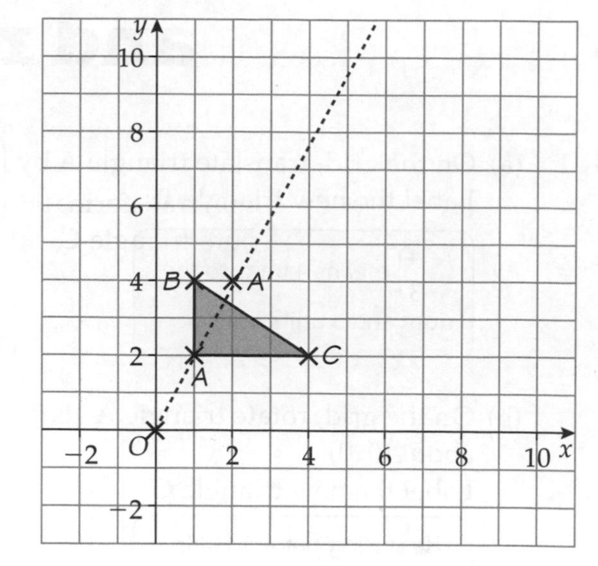

C **2** Describe fully the single transformation that maps triangle **A** onto triangle **B**.

....

....

Join corresponding vertices to find the centre of enlargement.

(3 marks)

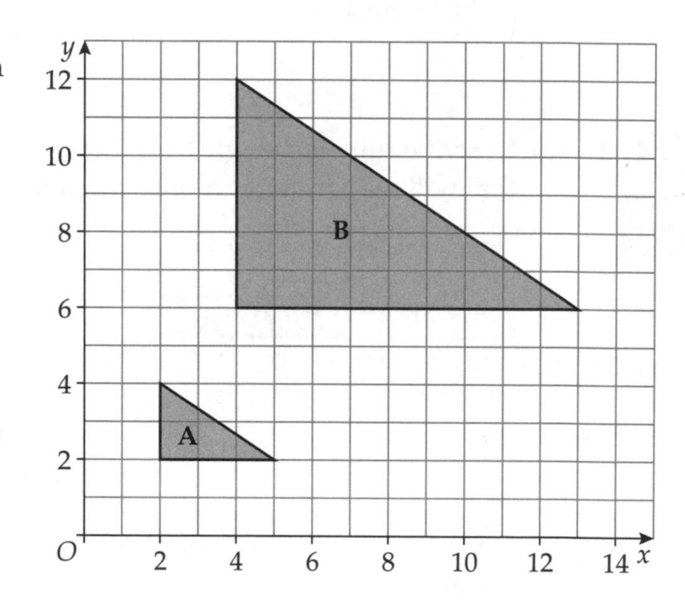

A* **3** (a) On the grid, enlarge triangle **C** by scale factor $-\frac{1}{2}$, centre (0, 0). Label the new triangle **D**.

For a negative enlargement, the image will be upside down on the other side of the centre of enlargement.

(3 marks)

(b) On the grid, enlarge triangle **C** by scale factor −2, centre (4, 4). Label the new triangle **E**.

(3 marks)

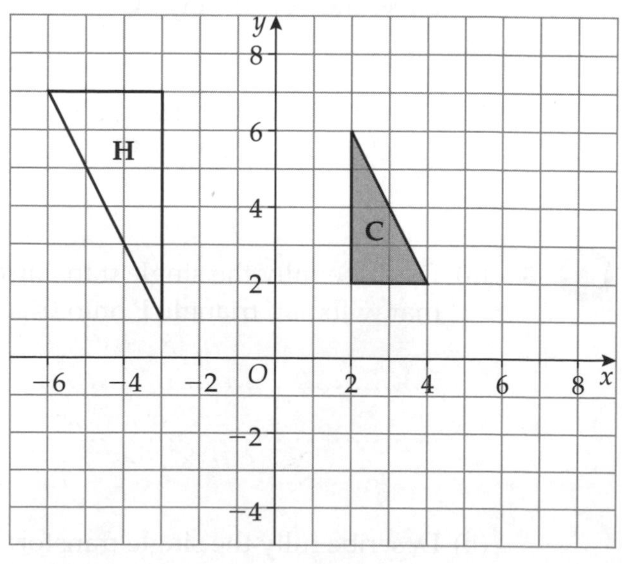

(c) Describe fully the single transformation that will map triangle **C** onto triangle **H**.

....

....

(3 marks)

Combining transformations

B **1** Triangle **A** is reflected in the line $x = 4$ to give triangle **B**.
Triangle **B** is reflected in the x-axis to give triangle **C**.
Describe fully the single transformation that will map triangle **A** onto triangle **C**.

....

....

....

(3 marks)

First carry out the two transformations on the diagram.

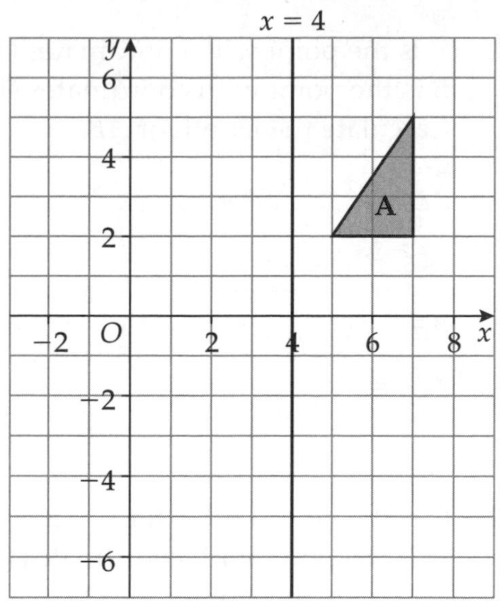

B **2** Shape **D** is reflected in the line $y = -3$ to give shape **E**.
Shape **E** is reflected in the x-axis to give shape **F**.
Describe fully the single transformation that will map shape **D** onto shape **F**.

....

....

....

(3 marks)

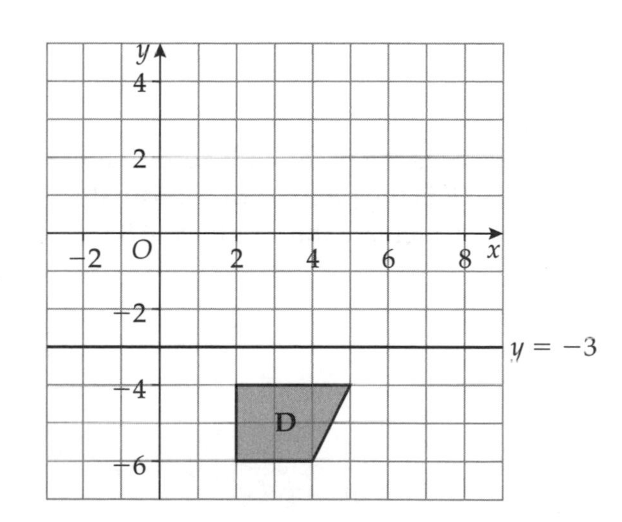

A* **3** Triangle **P** is enlarged by scale factor 2, centre (0, 0) to give triangle **Q**.
Triangle **Q** is rotated 180° about (0, 0) to give triangle **R**.
Describe fully the single transformation that will map triangle **P** onto triangle **R**.

....

....

....

(3 marks)

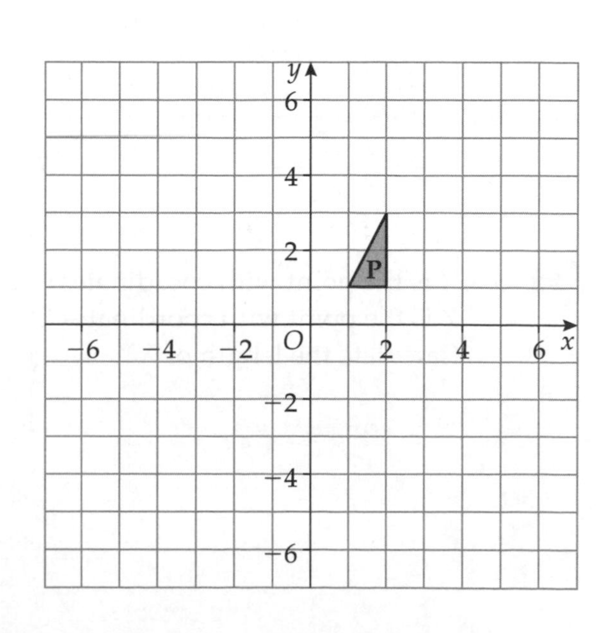

Line segments

D **1** A is the point with coordinates (1, 2).
B is the point with coordinates (13, 7).
Calculate the length of AB.

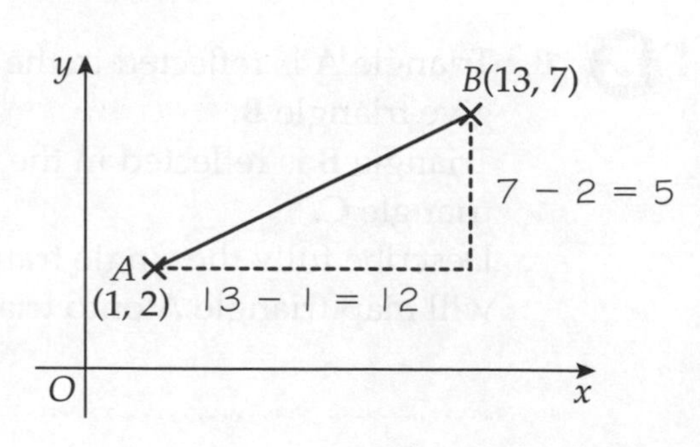

Guided

AB^2 = ...^{2} + ...^{2}

AB^2 = ...

AB = $\sqrt{.........}$

AB = ...

(3 marks)

D **2** C is the point with coordinates (5, 9).
D is the point with coordinates (2, 4).
Calculate the length of CD.

Sketch a diagram like that in question 1.

CD = ... **(3 marks)**

D **3** A is the point with coordinates (−8, −3).
B is the point with coordinates (4, 6).
Calculate the length of AB.

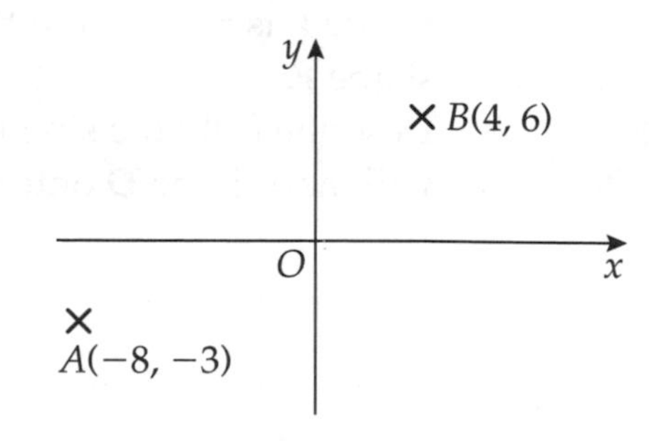

AB = ... **(3 marks)**

C **4** L is the point with coordinates (−9, 10).
M is the point with coordinates (1, −14).
Calculate the length of LM.

LM = ... **(3 marks)**

C **5** X is the point with coordinates (−3, −6).
Y is the point with coordinates (7, −1).
Calculate the length of XY.

XY = ... **(4 marks)**

Trigonometry 1

B 1 Work out the value of x.
Give your answer correct to 1 decimal place.

x
adjacent 4.5 cm
hypotenuse 8.7 cm

Guided

$\cos x = \frac{\text{adj}}{\text{hyp}}$

$\cos x = \frac{\dots\dots\dots}{\dots\dots\dots}$

$x = \cos^{-1}\left(\frac{\dots\dots\dots}{\dots\dots\dots}\right)$

$x = \dots\dots\dots\dots\dots\dots\dots$

$x = \dots\dots\dots\dots\dots\dots\dots°$ correct to 1 d.p.

> Start by labelling the sides of the triangle that you have been given. Then write down the trig ratio that uses these two sides.

(3 marks)

B 2 Calculate the size of angle y in this right-angled triangle.
Give your answer correct to 3 significant figures.

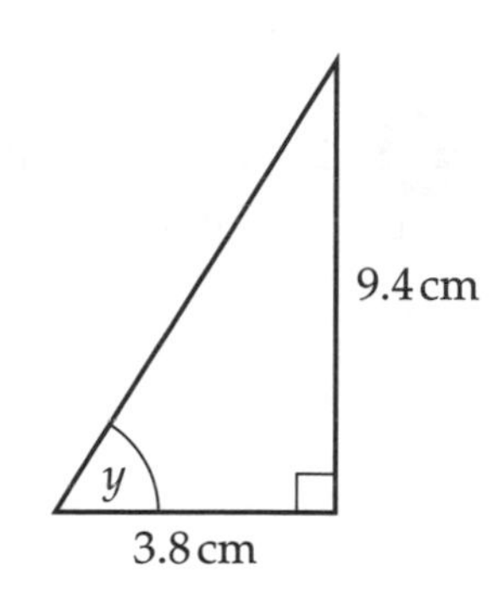

$y = \dots\dots\dots\dots\dots\dots\dots°$ **(3 marks)**

B 3 Work out the value of x.
Give your answer correct to 1 decimal place.

> First use triangle *DAB* and Pythagoras' theorem to work out the length of *BD*.

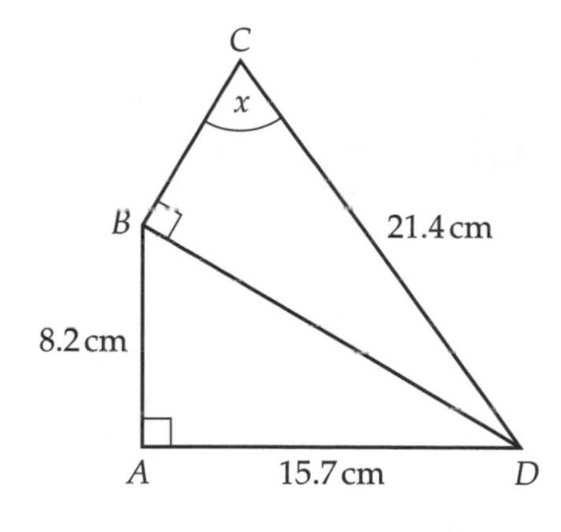

$x = \dots\dots\dots\dots\dots\dots\dots°$ **(5 marks)**

B 4 A ladder is 3.5 m long.
It leans against a vertical wall when placed on horizontal ground.
The ladder rests at a point on the wall that is 3 m above the ground.
What angle will the ladder make with the horizontal?

> Start by sketching a right-angled triangle to represent the ground, wall and ladder. Mark the known lengths on the triangle.

$\dots\dots\dots\dots\dots\dots\dots°$ **(3 marks)**

Trigonometry 2

B **1** Work out the length of AC.
Give your answer correct to 3 significant figures.

Start by labelling the side of the triangle that you have been given and the side that you want to find. Then write down the trig ratio that uses these two sides.

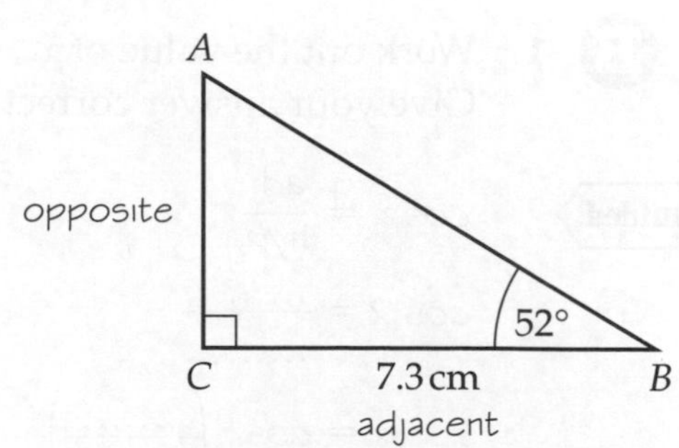

Guided

$\tan x = \frac{\text{opp}}{\text{adj}}$

$\tan(\ldots\ldots\ldots) = \frac{AC}{\ldots\ldots\ldots}$

$AC = \ldots\ldots\ldots \times \tan(\ldots\ldots\ldots)$

$AC = \ldots\ldots\ldots\ldots\ldots\ldots\ldots$

$AC = \ldots\ldots\ldots\ldots\ldots\ldots\ldots$ cm correct to 3 s.f. **(3 marks)**

B **2** Calculate the value of y.
Give your answer correct to 1 decimal place.

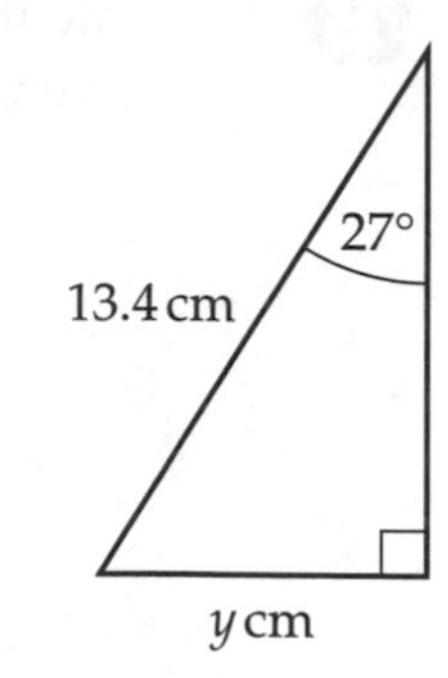

$y = \ldots\ldots\ldots\ldots\ldots\ldots\ldots$ cm **(3 marks)**

B **3** Work out the length of PR.
Give your answer correct to 3 significant figures.

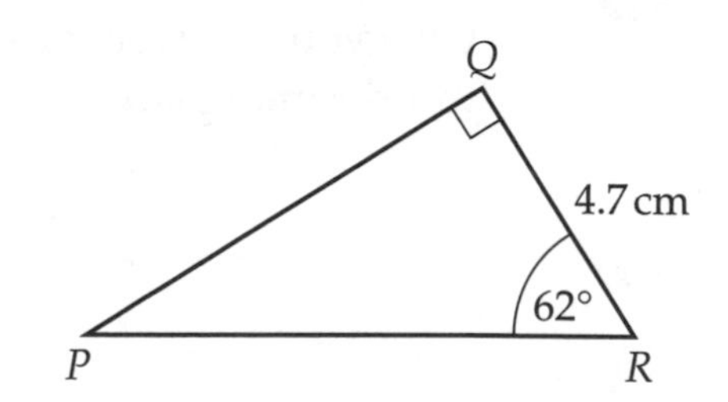

$PR = \ldots\ldots\ldots\ldots\ldots\ldots\ldots$ cm **(3 marks)**

A **4** The diagram shows two right-angled triangles.
Calculate the length of NP.
Give your answer correct to 3 significant figures.

First use triangle *LMP* to find the length of *MP*.

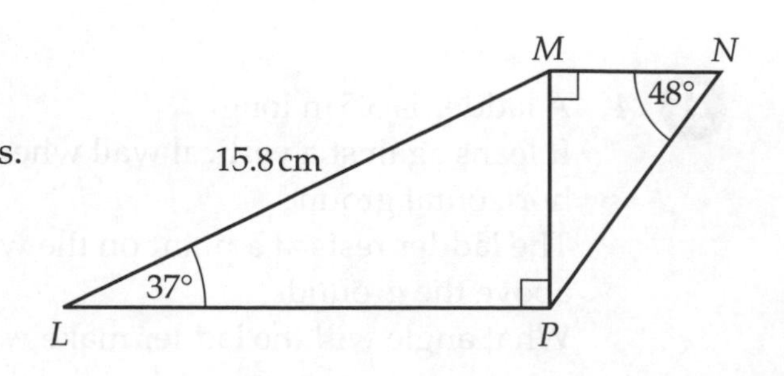

$NP = \ldots\ldots\ldots\ldots\ldots\ldots\ldots$ cm **(4 marks)**

Pythagoras in 3-D

1 A cuboid has length 8 cm, width 6 cm and height 15 cm.
Work out the length of PQ.
Give your answer correct to 3 significant figures.

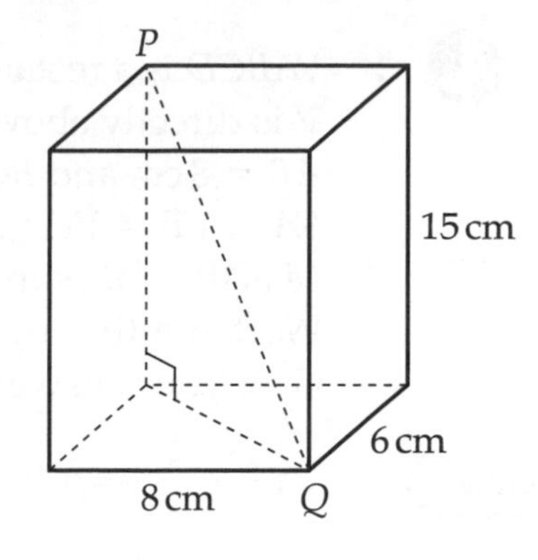

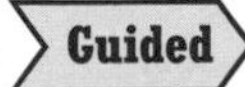

$PQ^2 = 8^2 + 6^2 + 15^2$

$= \dots\dots\dots\dots$

$PQ = \sqrt{\dots\dots}$

$= \dots\dots\dots\dots$

$= \dots\dots\dots\dots$ cm correct to 3 s.f.

(3 marks)

2 A cuboid has length 24 cm, width 7 cm and height 12 cm.
Show that a rod of length 28 cm will not fit completely in the cuboid.

Work out the length of the space diagonal of the cuboid.

(4 marks)

3 $ABCDEF$ is a triangular prism.
Work out the length of CE.
Give your answer correct to 3 significant figures.

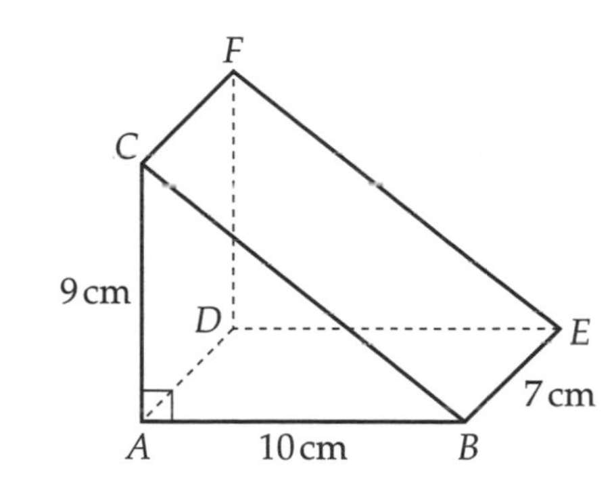

$CE = \dots\dots\dots\dots$ cm **(3 marks)**

4 V is directly above the centre of the rectangle.
$AB = 9$ cm and $BC = 6$ cm
$VA = VB = VC = VD = 11$ cm
Work out the height of the pyramid.
Give your answer correct to 3 significant figures.

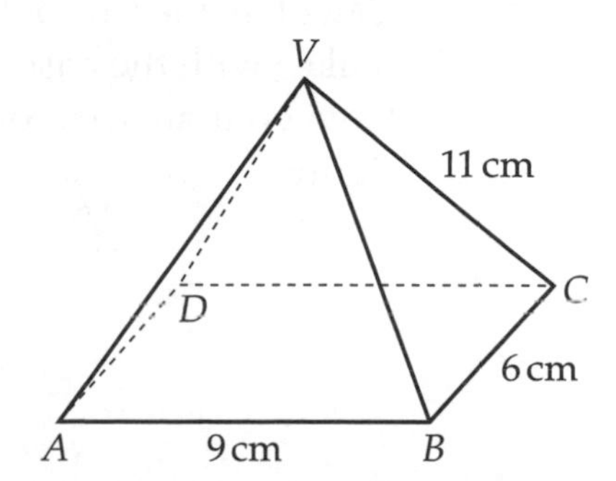

$\dots\dots\dots\dots$ cm **(3 marks)**

Trigonometry in 3-D

A* **1** *VABCD* is a rectangular-based pyramid.
V is directly above *X*, the centre of the rectangle.
$AB = 8$ cm and $BC = 6$ cm
$VA = VB = VC = VD = 12$ cm
M is the midpoint of *BC*.
Work out the angle that *VM* makes with the base.
Give your answer correct to 1 decimal place.

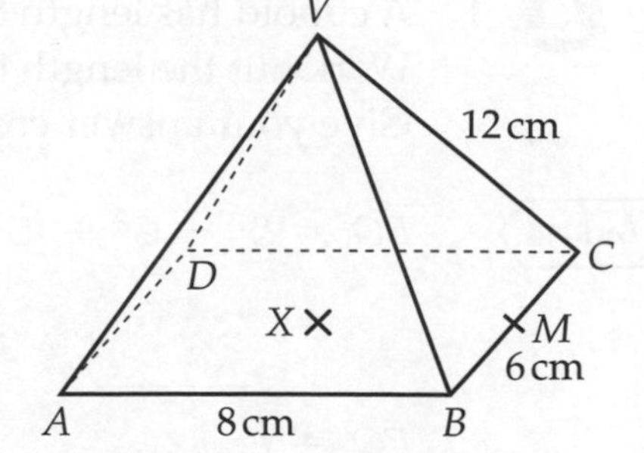

$VM^2 + 3^2 = 12^2$

$VM^2 =$ .... − ....

$VM = \sqrt{..............}$

$\cos VMX = \frac{.........}{.........}$

$VMX = \cos^{-1}\left(\frac{.........}{.........}\right)$

= ....

= ....° correct to 1 d.p.

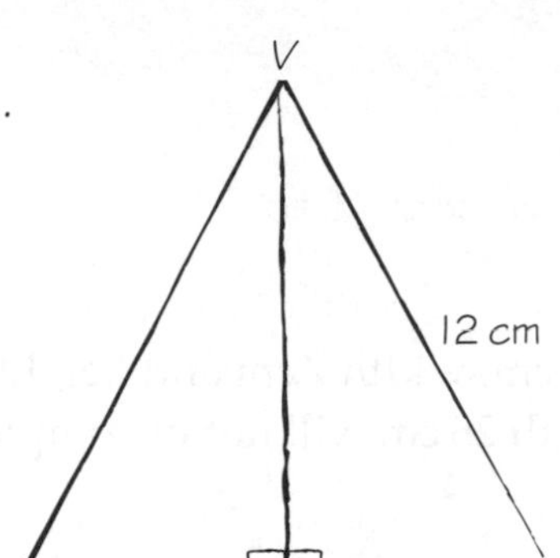

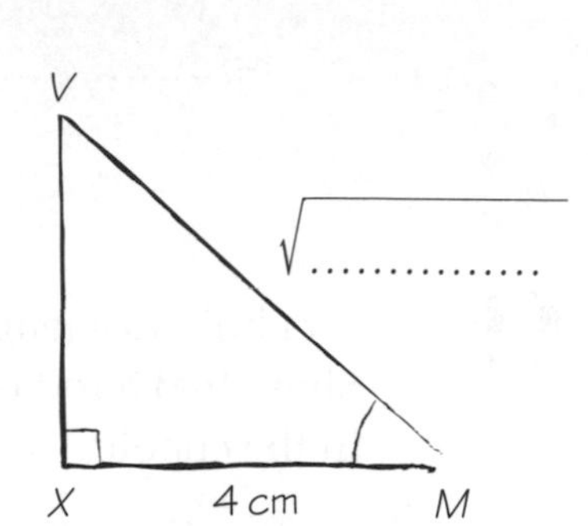

(4 marks)

A* **2** *ABCDEFGH* is a cuboid.
Calculate the angle that *HB* makes with the base.
Give your answer correct to 1 decimal place.

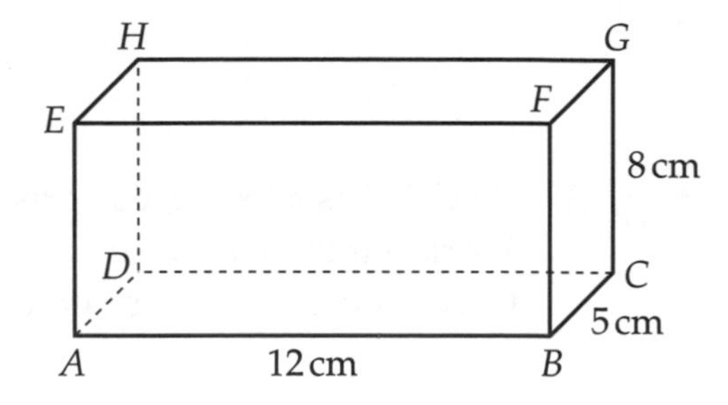

...................° **(4 marks)**

A* **3** The diagram shows a triangular prism.
Calculate the size of the angle that line *AF* makes with the base *ABCD*.
Give your answer correct to 3 significant figures.

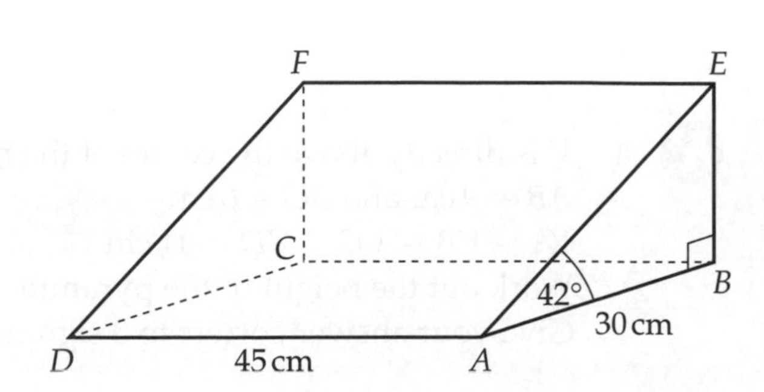

...................° **(4 marks)**

Triangles and segments

A 1 Calculate the area of triangle PQR.
Give your answer correct to 3 significant figures.

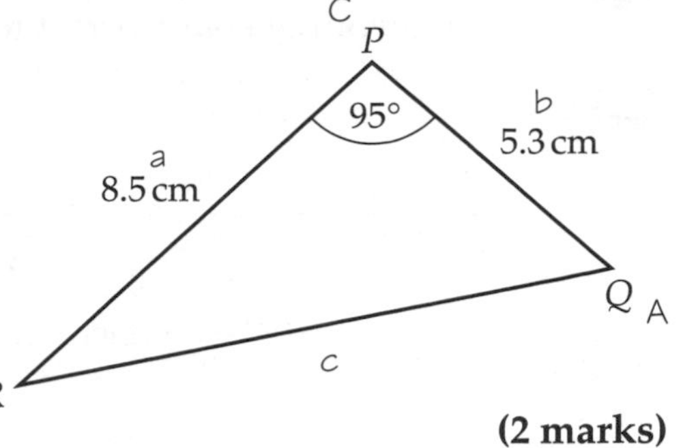

Guided

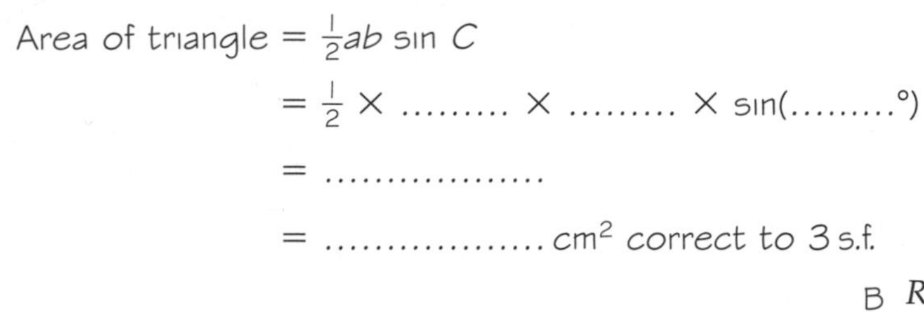

(2 marks)

A 2 Calculate the area of triangle LMN.
Give your answer correct to 3 significant figures.

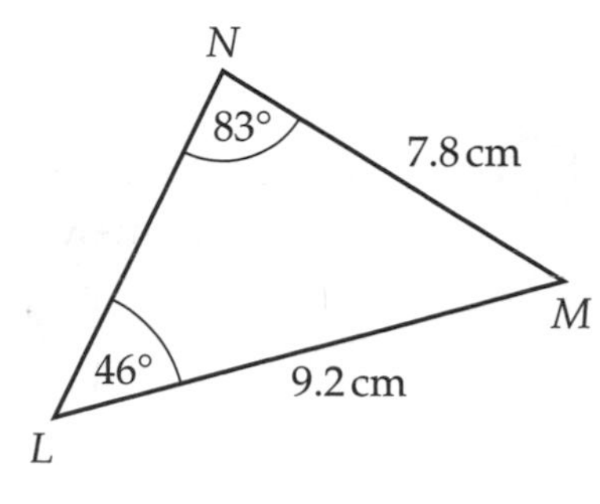

.... cm² **(3 marks)**

A 3 The area of triangle ABC is 40.4 m².
Work out the size of angle C.
Give your answer correct to 1 decimal place.

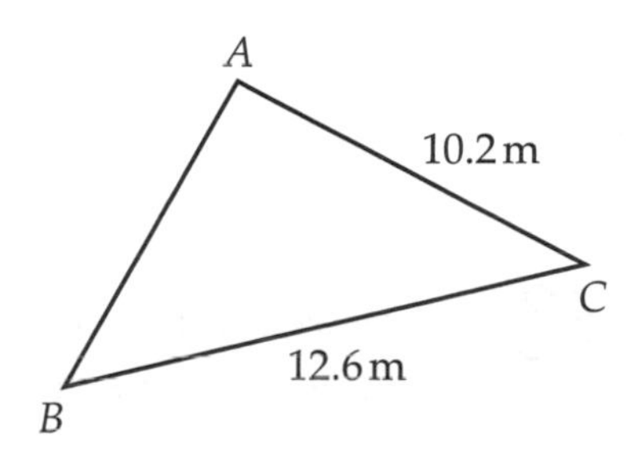

Guided

Area of triangle ABC = $\frac{1}{2}ab \sin C$

.... = $\frac{1}{2}$ × .... × .... × $\sin C$

$\frac{.........}{.........}$ = $\sin C$

$C = \sin^{-1}\left(\frac{.........}{.........}\right)$

C = ...

C = ...° correct to 1 d.p.

(3 marks)

A* 4 The diagram shows a sector of a circle with centre O.
$OA = OC = 15$ cm. Angle $AOC = 100°$.
Calculate the area of the shaded segment ABC.
Give your answer correct to 3 significant figures.

Area of segment = area of sector – area of triangle

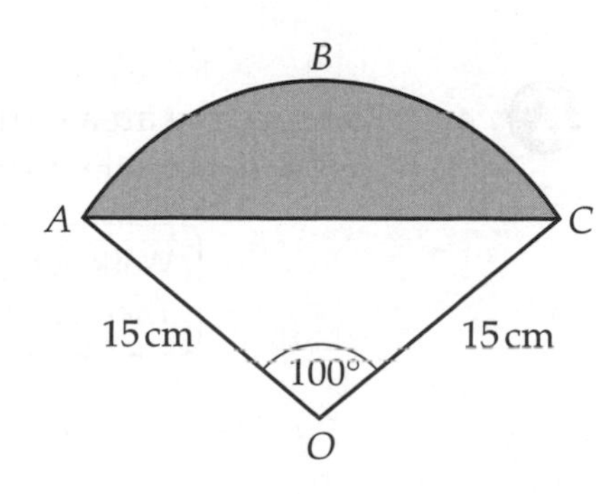

.... cm² **(3 marks)**

The sine rule

A **1** Work out the length of LM.
Give your answer correct to 3 significant figures.

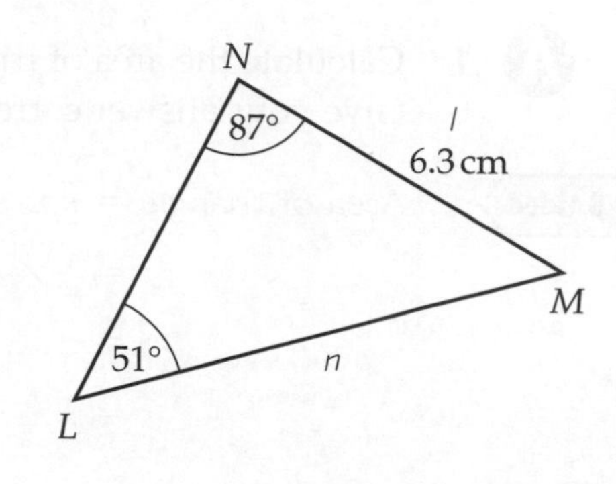

Guided

$$\frac{n}{\sin N} = \frac{l}{\sin L}$$

$$\frac{n}{\sin(\dots\dots\dots°)} = \frac{\dots\dots\dots}{\sin(\dots\dots\dots°)}$$

$$n = \frac{\dots\dots\dots}{\sin(\dots\dots\dots°)} \times \sin(\dots\dots\dots°)$$

$n = \dots\dots\dots\dots\dots\dots\dots$

$n = \dots\dots\dots\dots\dots\dots\dots$ cm correct to 3 s.f.

(3 marks)

A **2** Work out the size of angle ABC.
Give your answer correct to 1 decimal place.

First find angle C. Use $\frac{\sin C}{c} = \frac{\sin A}{a}$

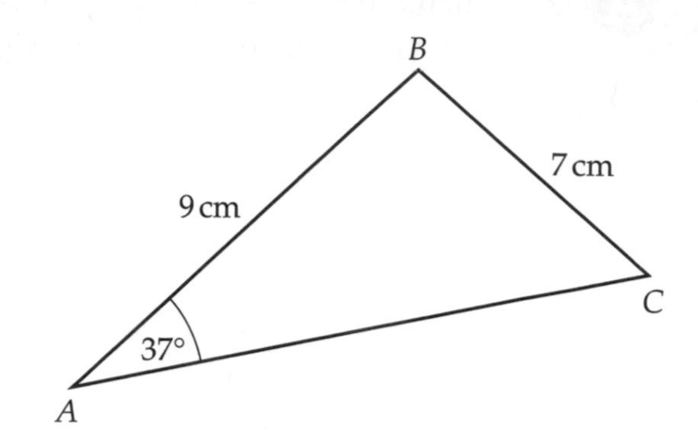

.....................° **(3 marks)**

A **3** Work out the perimeter of triangle FGH.
Give your answer correct to 3 significant figures.

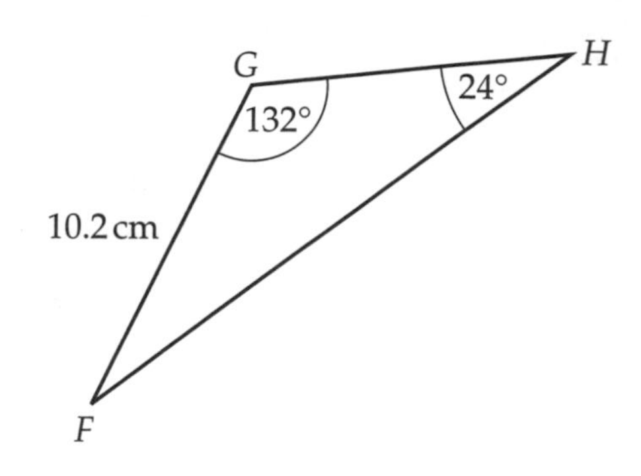

.... cm **(5 marks)**

A* **4** Work out the area of triangle PQR.
Give your answer correct to 3 significant figures.

Work out one of the missing lengths.
Then use area of triangle $= \frac{1}{2}ab \sin C$.

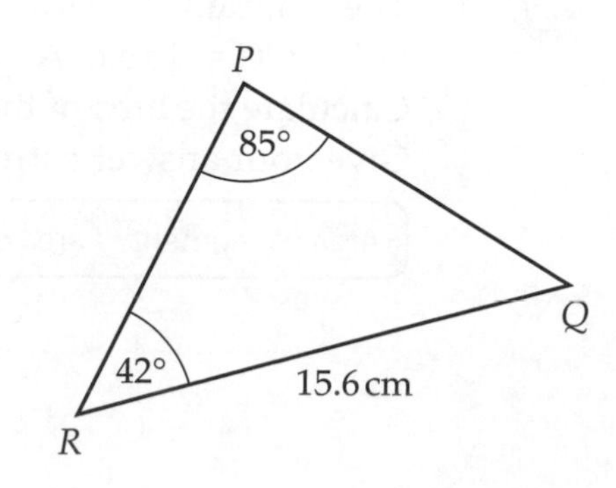

.... cm^2 **(5 marks)**

The cosine rule

A **1** Calculate the length of PR.
Give your answer correct to 3 significant figures.

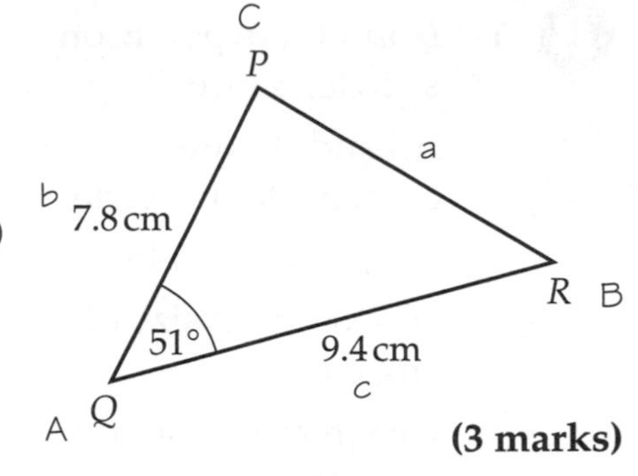

Guided

$a^2 = b^2 + c^2 - 2bc\cos A$

$PR^2 = (......)^2 + (......)^2 - 2 \times ...... \times ...... \times \cos(......°)$

$PR^2 =$ .... − ....

$PR =$ ....

$PR =$ .... correct to 3 s.f.

(3 marks)

A **2** (a) Work out the size of angle DEF.
Give your answer correct to 1 decimal place.

Rename the triangle ABC, replacing E with A as this is the angle to be found.

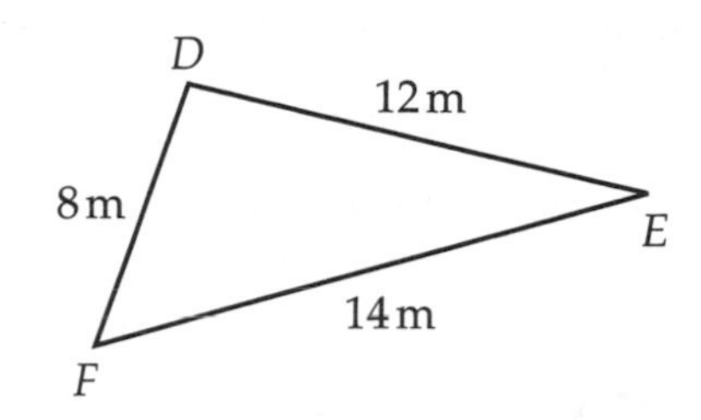

....................° **(3 marks)**

(b) Work out the area of triangle DEF.
Give your answer correct to 3 significant figures.

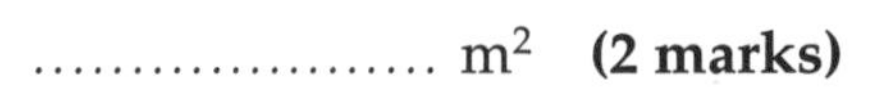

.... m² **(2 marks)**

A **3** Work out the perimeter of triangle ABC.
Give your answer correct to 3 significant figures.

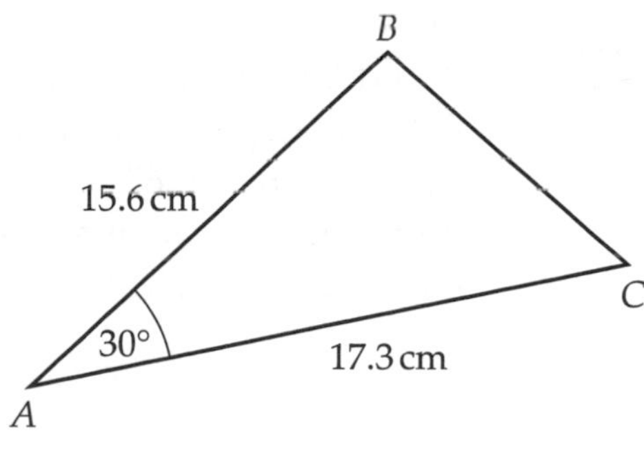

.... cm **(4 marks)**

A* **4** $ABCD$ is a quadrilateral.
Angle $BAC = 37°$.
Work out the area of $ABCD$.
Give your answer correct to 3 significant figures.

Draw in the line AC and mark in the size of angle BAC.

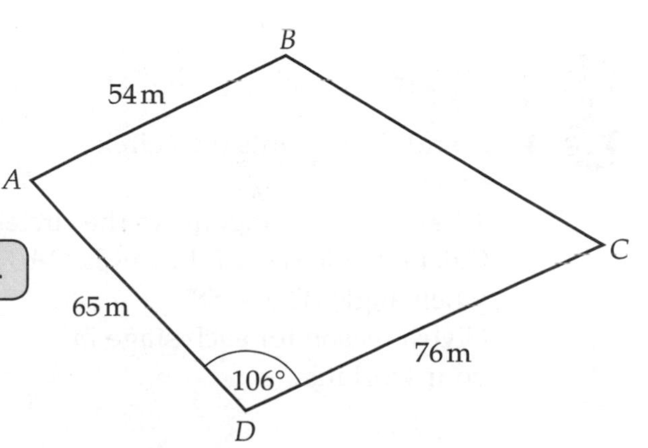

.... m² **(6 marks)**

Circle facts

B 1 B and C are points on a circle, centre O.
AB and AC are tangents to the circle.
Angle $BOC = 145°$.
Work out the size of angle BAO.
Give reasons for your answer.

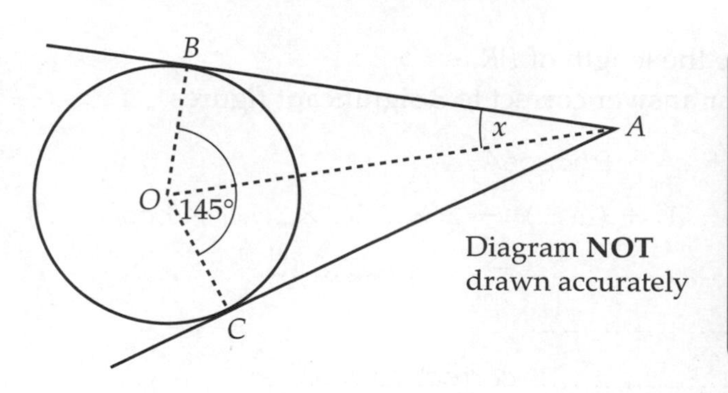

Make sure that all the information in the question is shown on the diagram. Also mark the angle you need to find.

Guided

Angle OBA = angle OCA = ...° Angle between tangent and radius is ....°

Angle CAB = 360° − 145° − ....° − ....° Angles in a quadrilateral add up to ....°

Angle BAO = ...° ÷ 2 Tangents from a point outside a circle are the same length.

= ....°

(4 marks)

B 2 R and S are points on a circle, centre O.
TR and TS are tangents to the circle.
Angle $ORS = 23°$.
Work out the size of angle RTS.
Give reasons for your answer.

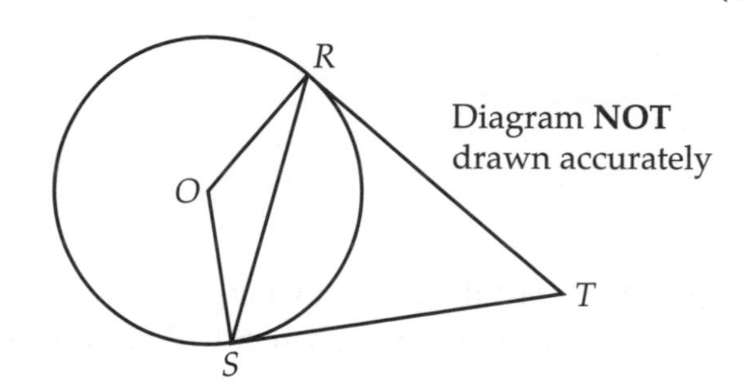

.....................° **(4 marks)**

B 3 A and B are points on a circle, centre O.
CA is a tangent to the circle.
Angle $AOB = 126°$.
Work out the size of angle x.
Give reasons for your answer.

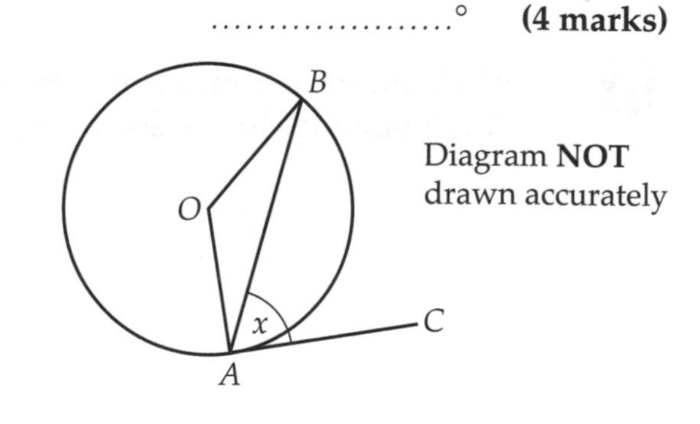

.....................° **(4 marks)**

A 4 A and B are points on a circle, centre O.
TA and TB are tangents to the circle.
Calculate the size of the angle OAB when angle $ATB = 58°$.
Give a reason for each stage in your working.

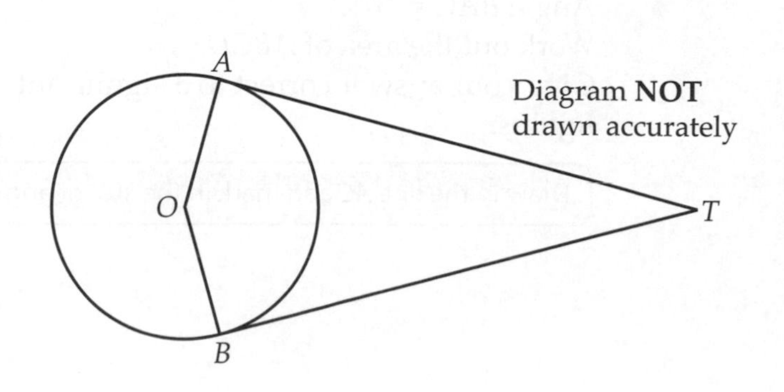

.....................° **(4 marks)**

Circle theorems

1 In the diagram, A, B and C are points on the circumference of a circle, centre O.
Angle $ABC = 75°$.
Work out the size of angle OCA.
Give reasons for your answer.

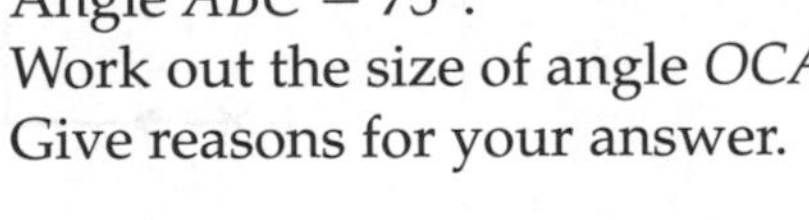

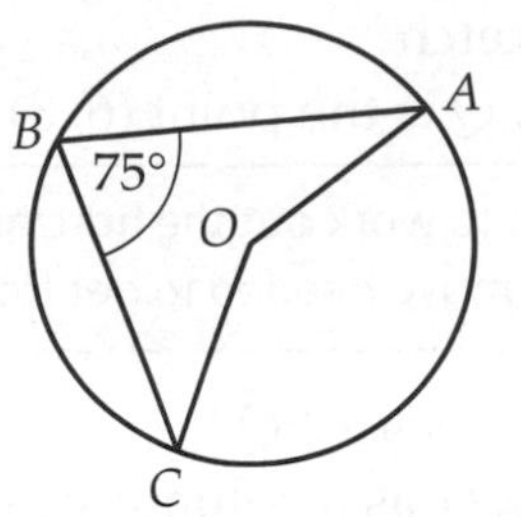

Exam questions similar to this have proved especially tricky – be prepared! **ResultsPlus**

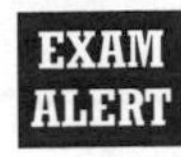

Working | Reason

Angle $COA = 2 \times 75°$

$= \ldots\ldots\ldots°$ ....

Angle OCA = angle OAC ....

Angle $OCA = \dfrac{180° - \ldots\ldots\ldots}{2}$ ....

$= \ldots\ldots\ldots\ldots\ldots\ldots\ldots°$ **(3 marks)**

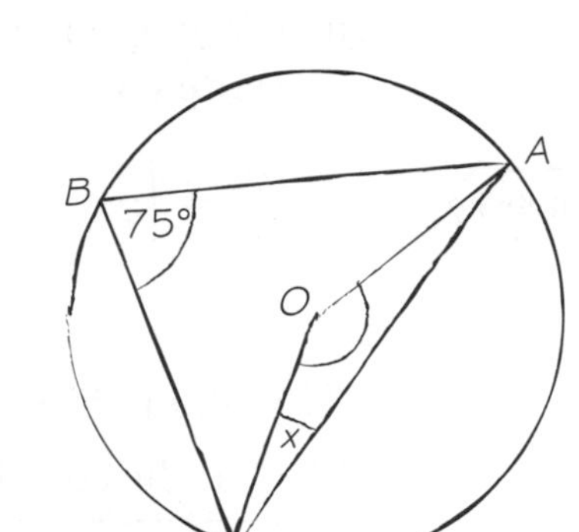

A **2** A, B, C and D are points on the circumference of a circle, centre O.
Angle $ADC = 114°$.
Work out the size of the angle marked x.
Give reasons for your answer.

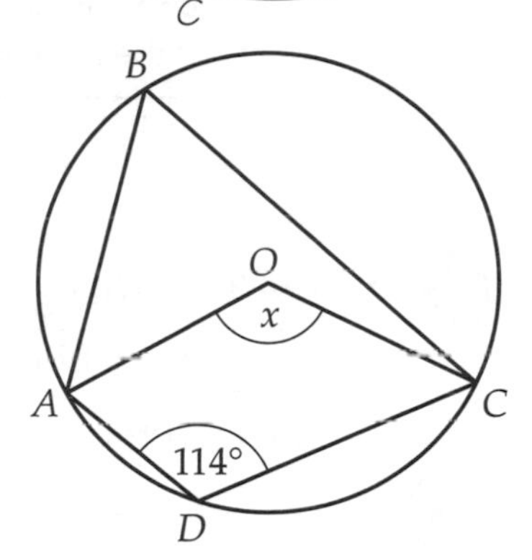

.....................° **(4 marks)**

3 P, Q, R and S are points on the circumference of a circle.
Angle $RPQ = 50°$.
RP is a diameter.
Work out the size of angle QSP.
Give reasons for your answer.

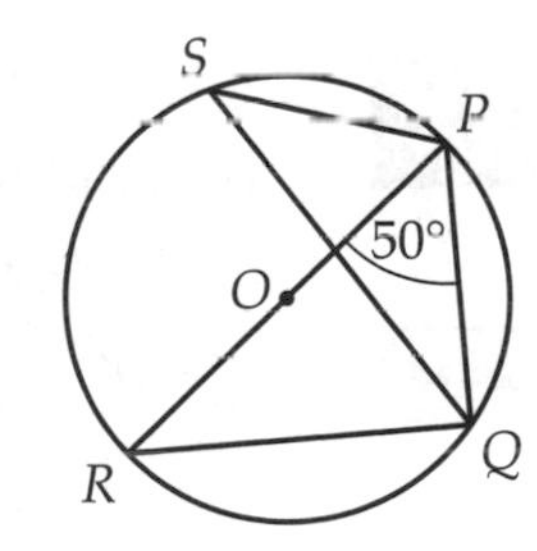

.....................° **(4 marks)**

A* **4** D, E and F are points on the circumference of a circle.
GD is a tangent to the circle.
Work out the size of angle EDF.
Give reasons for your answer.

Use the alternate segment theorem.

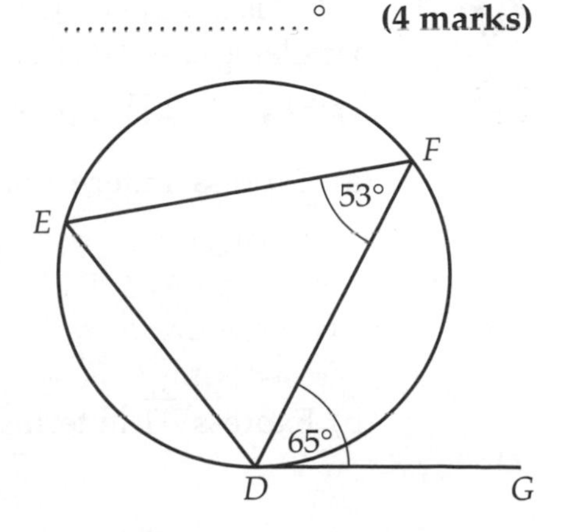

.....................° **(4 marks)**

Vectors

1 The diagram is a sketch.
P is the point (3, 4). Q is the point (6, 8).

> Use the diagram to work out the horizontal move and the vertical move needed to get from P to Q.

(a) Write down the vector $\overrightarrow{PQ}$.
Write your answer as a column vector $\begin{pmatrix} x \\ y \end{pmatrix}$.

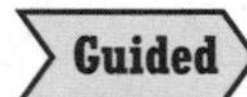

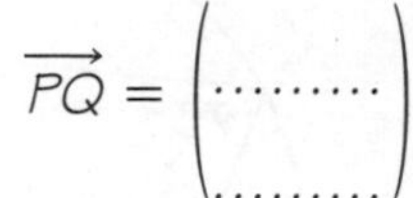

$\overrightarrow{PQ} = \begin{pmatrix} \ldots\ldots\ldots \\ \ldots\ldots\ldots \end{pmatrix}$

(2 marks)

$PQRS$ is a parallelogram. $\overrightarrow{PS} = \begin{pmatrix} 4 \\ 1 \end{pmatrix}$

(b) Find the coordinates of the point R.

> $PQRS$ is a parallelogram so draw this on the diagram. QR is parallel to PS.

.... **(2 marks)**

2 OAB is a triangle.
$\overrightarrow{OA} = \mathbf{a}$ $\quad \overrightarrow{OB} = \mathbf{b}$

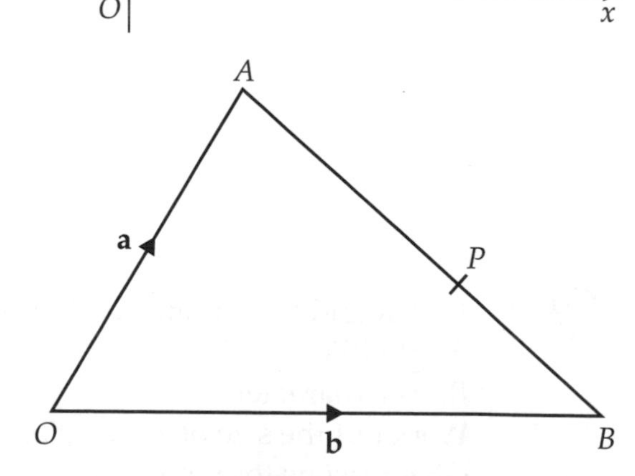

(a) Find the vector $\overrightarrow{AB}$ in terms of **a** and **b**.

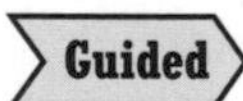

$\overrightarrow{AB} = \overrightarrow{AO} + \overrightarrow{OB}$

$\overrightarrow{AB} = \ldots\ldots\ldots\ldots\ldots\ldots$ **(1 mark)**

P is the point on AB such that $AP : PB = 3 : 2$

(b) Find the vector $\overrightarrow{OP}$ in terms of **a** and **b**.
Give your answer in its simplest form.

$\overrightarrow{OP} = \overrightarrow{OA} + \overrightarrow{AP}$

$= \overrightarrow{OA} + \frac{3}{5}\overrightarrow{AB}$

$= \ldots\ldots\ldots\ldots\ldots\ldots + \frac{3}{5}(\ldots\ldots\ldots\ldots\ldots\ldots)$

$= \ldots\ldots\ldots\ldots\ldots\ldots$

$= \ldots\ldots\ldots\ldots\ldots\ldots$

> Exam questions similar to this have proved especially tricky – be prepared! ResultsPlus

> Simplify your answer.

(3 marks)

3 $ABCD$ is a parallelogram. AC and BD are diagonals of parallelogram $ABCD$. AC and BD intersect at T.
$\overrightarrow{AB} = \mathbf{a}$ $\quad \overrightarrow{AD} = \mathbf{d}$

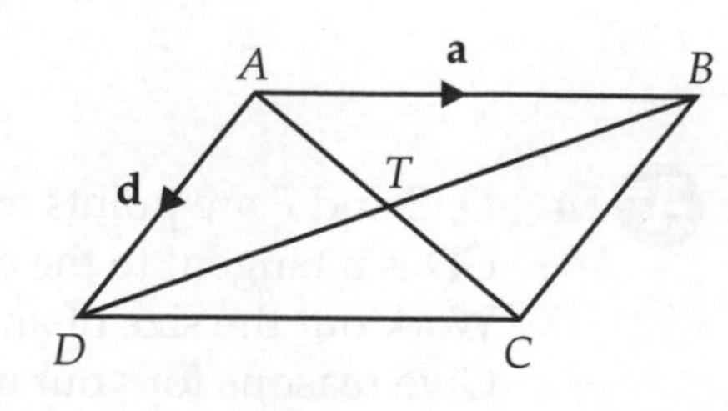

(a) Express, in terms of **a** and **d**

(i) $\overrightarrow{AC}$ $\quad$ (ii) $\overrightarrow{BD}$

$\overrightarrow{AC} = \ldots\ldots\ldots\ldots\ldots\ldots$

$\overrightarrow{BD} = \ldots\ldots\ldots\ldots\ldots\ldots$ **(2 marks)**

(b) Express $\overrightarrow{AT}$ in terms of **a** and **d**.

.... **(2 marks)**

Solving vector problems

1 *OPQR* is a trapezium with *PQ* parallel to *OR*.

$\overrightarrow{OP} = 4\mathbf{a}$ $\quad\overrightarrow{PQ} = 4\mathbf{b}$ $\quad\overrightarrow{OR} = 12\mathbf{b}$

M is the midpoint of *PQ* and *N* is the midpoint of *OR*.

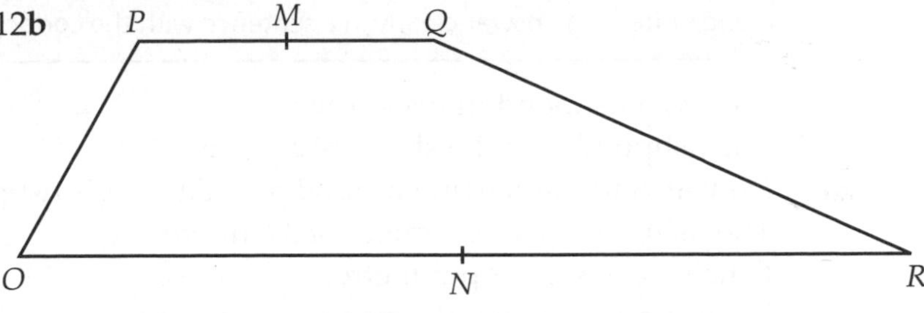

(a) Find the vector $\overrightarrow{MN}$ in terms of **a** and **b**.

Guided

$\overrightarrow{MN} = \overrightarrow{MP} + \overrightarrow{PO} + \overrightarrow{ON}$

$= \frac{1}{2}\overrightarrow{QP} + \overrightarrow{PO} + \frac{1}{2}\overrightarrow{OR}$

$=$ ....

$=$ ....

If $\overrightarrow{PQ} = 4\mathbf{b}$ then $\overrightarrow{QP} = -4\mathbf{b}$.

(2 marks)

X is the midpoint of *MN* and *Y* is the midpoint of *QR*.

(b) Prove that *XY* is parallel to *OR*.

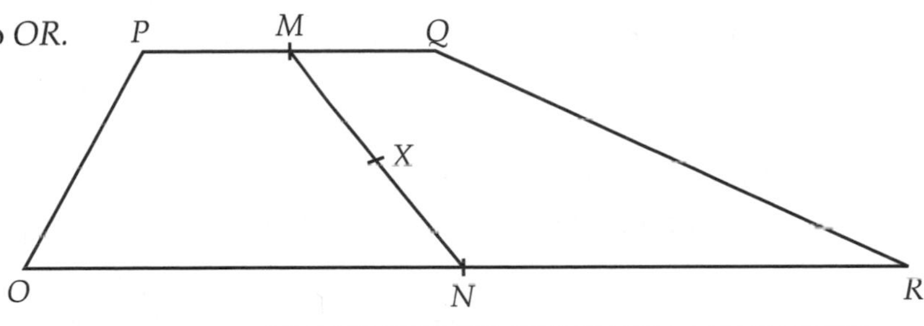

Find an expression for $\overrightarrow{XY}$ in terms of **a** and **b**. Then show that $\overrightarrow{XY}$ is also a multiple of **b**.

(2 marks)

2 *OAB* is a triangle.

B is the midpoint of *OR*. *Q* is the midpoint of *AB*.

$\overrightarrow{OA} = 3\mathbf{a}$ $\quad\overrightarrow{OB} = \mathbf{b}$

P divides *OA* in the ratio 2 : 1

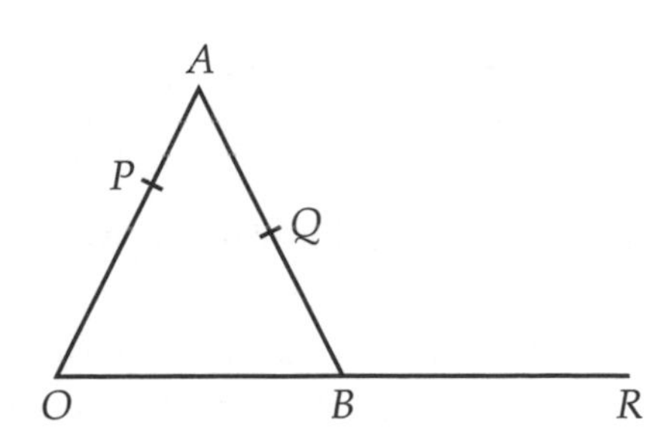

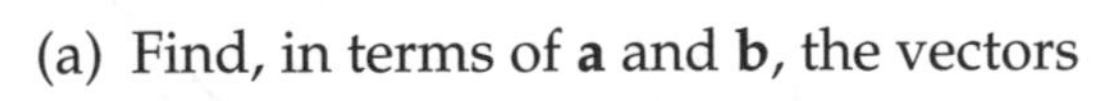

(a) Find, in terms of **a** and **b**, the vectors

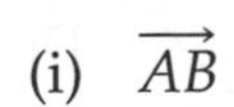

(i) $\overrightarrow{AB}$ .... **(1 mark)**

(ii) $\overrightarrow{PR}$

.... **(2 marks)**

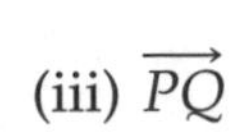

(iii) $\overrightarrow{PQ}$

.... **(2 marks)**

(b) Hence, explain why *PQR* is a straight line.

(2 marks)

Problem-solving practice

C *1

The question has a * next to it, so you must show all your working and write your answer clearly in a sentence with the correct units.

Jim has a fishpond in his garden.
The fishpond is in the shape of a circle.
He wants to put fencing around the edge of his fishpond.
The fishpond has a diameter of 3.6 metres.
Fencing costs £5.69 per metre.
Jim can only buy whole metres of fencing.
Work out the total cost of the fencing that Jim needs to buy.

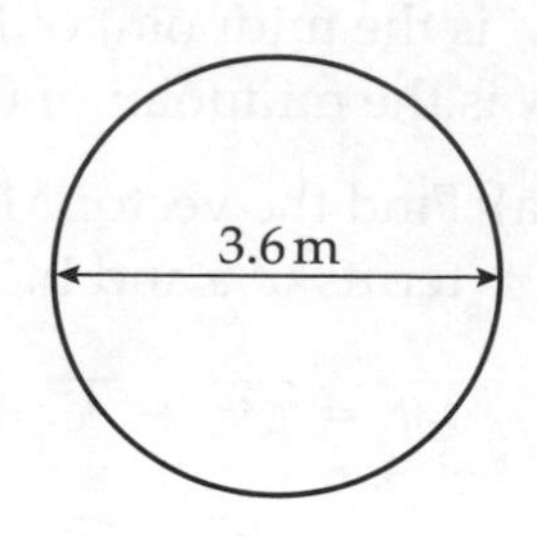

Guided

Circumference of pond = $\pi \times d$

= $\pi \times$ ....

= .... m (to 3 s.f.)

Whole metres of fencing needed = ... m

Cost of fencing = ... × ....

= £...........

(4 marks)

C 2 Describe fully the single transformation that will map triangle **A** onto triangle **B**.

There are 3 marks for the question so you should give 3 pieces of information.

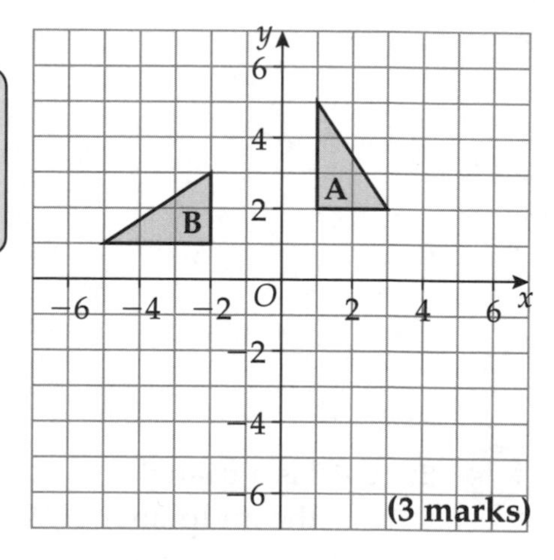

....

....

....

(3 marks)

B 3 The diagram shows two right-angled triangles.
$PL = 5.9$ cm. $LM = 4.6$ cm. $PN = 12.4$ cm.
Work out the size of angle PNM.
Give your answer correct to 1 decimal place.

First use Pythagoras' theorem in triangle *PML* to work out the length of *PM*. Then use trigonometry in triangle *PNM* to work out the size of angle *PNM*.

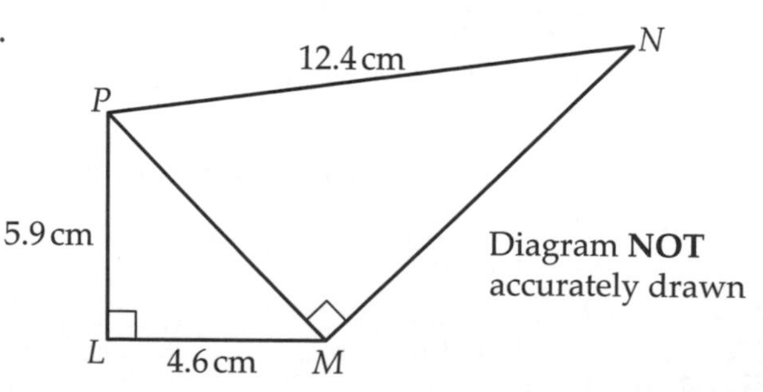

....................° **(5 marks)**

Problem-solving practice

 *4

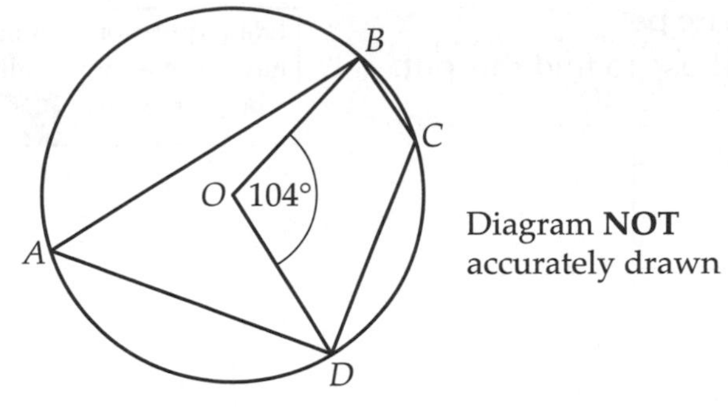

Diagram **NOT** accurately drawn

A, *B*, *C* and *D* are points on the circumference of a circle, centre *O*.
Angle $BOD = 104°$.
Work out the size of angle *BCD*.
Give a reason for each stage of your working.

Work out the size of angle *BAD* and then angle *BCD*. Make sure that you write down any circle theorem or angle fact that you use.

(4 marks)

 **5** The diagram shows a field.

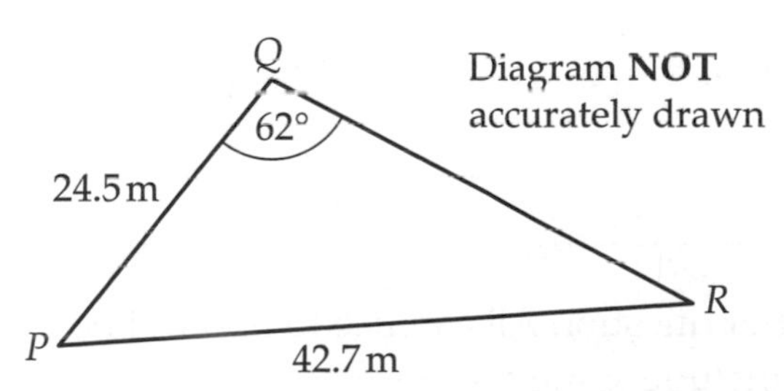

Diagram **NOT** accurately drawn

$PQ = 24.5$ m. $PR = 42.7$ m.
Angle $PQR = 62°$.
Work out the perimeter of the field.
Give your answer correct to 3 significant figures.

First use the sine rule to work out the size of angle *QRP*. Then work out angle *QPR* using the sum of the angles in a triangle. Use the sine rule or cosine rule to work out the length of *QR*. You can now work out the perimeter of the field.

....................m **(5 marks)**

Collecting data

D **1** Mary wants to find out people's favourite pet.
Design a data collection sheet she could use to find this out.

Exam questions similar to this have proved especially tricky – be prepared! ResultsPlus

Guided

EXAM ALERT

Pet		

(2 marks)

D **2** Katie wants to find out how often her friends go to the cinema.
She uses this question on a questionnaire.

How many times do you go to the cinema?		
☐ 1–2	☐ 2–5	☐ 6+

Make sure that you find **different** things that are wrong – look at both the question and the response boxes.

Write down **two** things wrong with this question.

1 ....

2 ....

(2 marks)

C **3** Ali wants to find out how many emails people receive.
She uses this question on a questionnaire.

How many emails do you receive?		
☐ Not many	☐ Quite a few	☐ A lot

(a) Write down **two** things wrong with this question.

1 ....

2 ....

(b) Design a better question Ali can use to find out how many emails people receive.
You should include some response boxes.

(2 marks)

C **4** Anthony wants to design a question for a questionnaire to find out how often people go on holiday. Design a suitable question he could use.

Remember to include a time frame and some response boxes with your question.

(2 marks)

Two-way tables

D 1 80 British students each visited one foreign country last week.
The two-way table shows some information about these students.

	France	Germany	Spain	Total
Female			13	38
Male	21			
Total		25	22	80

Work out the missing values in this row first.
Start with the total column.

Complete the two-way table.

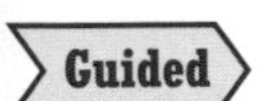

'Total' column: $80 - 38 = 42$

(3 marks)

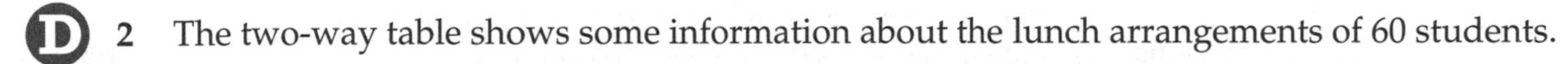

D 2 The two-way table shows some information about the lunch arrangements of 60 students.

	School lunch	Packed lunch	Other	Total
Female	16			39
Male	9	8		
Total			21	60

Complete the two-way table. **(3 marks)**

C 3 40 students were asked if they liked coffee.
25 of the students were girls.
13 boys liked coffee.
14 girls did **not** like coffee.
Use this information to complete the two-way table.

Transfer the information from the question to the two-way table first. Then complete the table.

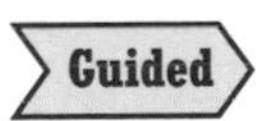

	Boys	Girls	Total
Liked coffee	....		
Did not like coffee		....	
Total		....	40

(3 marks)

C 4 75 adults were asked which type of music they preferred.
47 of the adults were male.
2 of the females and 8 of the males preferred country.
32 of the adults preferred jazz.
9 of the males preferred other types of music.
Draw up a two-way table to show this information.

(4 marks)

Stratified sampling

1 The table shows the number of students in each year group at a school.

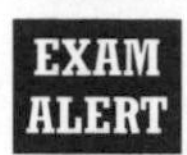

Year group	7	8	9	10	11
Number of students	210	172	151	160	148

Exam questions similar to this have proved especially tricky – be prepared! ResultsPlus

Harry is carrying out a survey of these students.
He uses a stratified sample of 80 students according to year group.
Calculate the number of Year 9 students that should be in his sample.

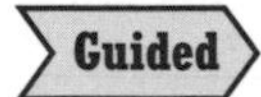

Total number of students = 210 + 172 + ... + ... + ...

= ....

Number of Year 9 in sample = $\frac{\dots\dots}{\dots\dots} \times 80$

Write down the fraction of the students that are in Year 9.

= ....

= .... students

Write down at least five figures from your calculator display and then round to the nearest whole number.

(3 marks)

2 In a school 277 students each study one of three languages.

	German	French	Spanish
Male	32	49	27
Female	41	56	72

A sample, stratified by the language studied and by gender, of 60 of the 277 students is taken.
Work out the number of female students studying French in the sample.

.... **(2 marks)**

3 The table shows the number of students in each of three schools.

School	A	B	C
Number of students	780	908	1045

Jerry takes a sample of these students stratified by school.
There are 46 students from school C in his sample.
Work out the number of students from school A in his sample.

Guided

Fraction of students in sample = $\frac{x}{780} = \frac{\dots\dots}{1045}$

The fraction of students in each school will be the same.

$x = \frac{\dots\dots}{1045} \times 780$

$x =$ ....

Number from school A in sample = ...

(2 marks)

4 The table shows the number of employees in three different age groups.

Age	18–25 years	26–39 years	40+ years
Number of employees	95	376	189

A sample stratified by age group is taken.
There are 9 employees in the 18–25 age group.
Work out the number of employees in the 40+ age group in the sample.

.... **(3 marks)**

Mean, median and mode

C 1 Here are the numbers of texts received by 10 adults in one day.

2 4 7 6 25 24 16 5 4 3

(a) Write down the mode.

Guided The mode is ....

The mode is the value that appears most often.

(1 mark)

(b) Work out the median.

Guided 2 3 4 ....

First write out all the numbers in order.

Median = $\frac{\ldots\ldots + \ldots\ldots}{2}$

= ....

(2 marks)

(c) Work out the mean.

Add up all the numbers **before** dividing by 10.

Guided The total of all values = 2 + 4 + 7 + 6 + 25 + 24 + 16 + 5 + 4 + 3

= ....

Mean = $\frac{\ldots\ldots}{10}$

= ....

Parts (a) to (c) of this question are grade F or below. Part (d) is grade C and is the tricky bit. There is more than one possible answer to this, which is why you must give a reason.

(2 marks)

(d) Which of these three averages best describes this data? Give a reason for your answer.

Guided The average that best describes this data is .... because ....

.... **(1 mark)**

C 2 There are 6 boys and 2 girls in a room.
The mean of all their ages is 15
The mean of the girls' ages is 16.5
Work out the mean of the boys' ages.

Guided Total of the ages of all 8 children = 8 × 15 = ...

Total of the ages of the 2 girls = 2 × .... = ...

Total of boys' ages = ... − ....

= ....

Mean of boys' ages = $\frac{\ldots\ldots}{6}$

= ....

(3 marks)

C 3 The mean weekly wage of 5 women is £280
The mean weekly wage of 7 men is £316
Work out the mean weekly wage of all 12 people.

£..................... **(3 marks)**

Frequency table averages

D 1 The table shows information about the numbers of goals scored by a football team in their last 20 matches.

Guided

Number of goals	Frequency	Number of goals × frequency
0	8	0 × 8 = ...
1	6	1 × 6 = ...
2	3	.... × .... = ...
3	1	.... × .... = ...
4	2	.... × .... = ...

Add a 'Number of goals × frequency' column to the table to work out the total number of goals. If this column is not given, add it on yourself.

(a) Write down the mode.

The mode is the number of goals scored most often.

Guided The mode is .... goals. **(1 mark)**

(b) Find the median.

Guided The median will be the $\frac{20 + 1}{2}$th = ... th value.

The median is .... **(1 mark)**

(c) Work out the mean.

Add up the final column to work out the total number of goals.

.... goals **(3 marks)**

2 The table shows information about the numbers of minutes 40 children take to get to school.

Guided

Number of minutes (m)	Frequency	Midpoint	Midpoint × frequency
$0 < m \leqslant 10$	13	5	13 × 5 = ...
$10 < m \leqslant 20$	16	15	16 × 15 = ...
$20 < m \leqslant 30$	8	....	...... × ...... = ...
$30 < m \leqslant 40$	3	....	...... × ...... = ...

Exam questions similar to this have proved especially tricky – be prepared! ResultsPlus

The midpoint is the middle value of the class interval.
The midpoint is used as an estimate for the number of minutes.

(a) Write down the modal class.

.... **(1 mark)**

(b) Find the class interval that contains the median.

.... **(1 mark)**

(c) Work out an estimate for the mean number of minutes.

Guided Mean = $\frac{\text{total number of minutes}}{40}$ = $\frac{......}{40}$

Add up the final column to work out the total number of minutes.

= .... minutes **(4 marks)**

Interquartile range

B **1** Here are the numbers of text messages received by 15 people one day.

0 1 1 2 5 5 8 10 14 15 15 15 18 20 24

(a) Work out the range.

.... **(1 mark)**

(b) Work out the interquartile range.

Guided

Interquartile range = upper quartile – lower quartile

= .... – ....

= .... **(2 marks)**

B **2** The stem and leaf diagram shows the numbers of minutes 19 students took to complete their maths homework.

0	7 8 8 9
1	2 3 5 6
2	4 4 5 5 8
3	0 0 0 5 7 8

Use the key to interpret the stem and leaf diagram.

Key : 4 | 0 means 40 minutes

(a) Work out the range.

.... **(1 mark)**

(b) Work out the interquartile range.

.... **(2 marks)**

B **3** Jake counted the number of letters in each of 35 sentences in a newspaper.
He showed his results in a stem and leaf diagram.

0	6 6 7 8 8 9
1	1 2 3 4 4 8 9
2	0 3 5 5 7 7 8
3	2 2 3 3 6 6 8 8
4	1 2 3 3 5 6 7

Key : 4 | 1 means 41 letters

Work out the interquartile range.

.... **(2 marks)**

Frequency polygons

C 1 Emily carried out a survey of 70 students.
She asked them how many DVDs they each have.
This table gives information about the numbers of DVDs these students have.

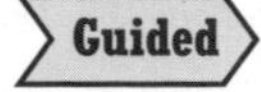

Midpoint	2	7	12	……	……
Number of DVDs	0–4	5–9	10–14	15–19	20–24
Frequency	6	19	15	23	7

On the grid, draw a frequency polygon to show this information.

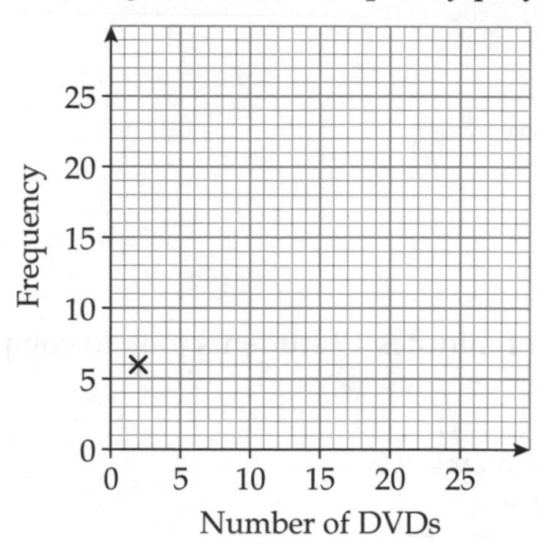

Plot the points at the midpoint of each class interval. Plot the second point at (7, 19).

(2 marks)

C 2 The table gives some information about the weights (w grams) of 60 apples.

Weight (w grams)	Frequency
$100 \leqslant w < 110$	5
$110 \leqslant w < 120$	9
$120 \leqslant w < 130$	14
$130 \leqslant w < 140$	24
$140 \leqslant w < 150$	8

On the grid, draw a frequency polygon to show this information.

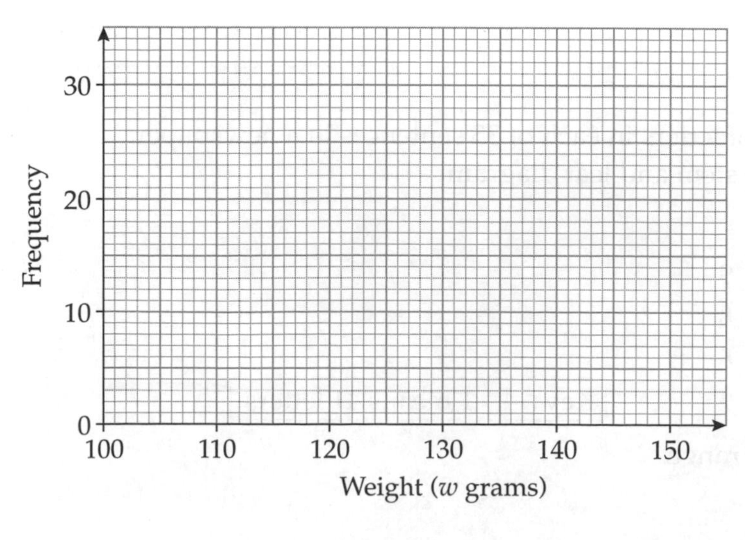

(2 marks)

Histograms

 **1** Sam asks some students how long they took to finish their science homework. The table and histogram show some of this information.

Guided

Time (minutes)	Frequency
$5 < x \leqslant 10$	... × 5 = ...
$10 < x \leqslant 15$	20
$15 < x \leqslant 25$	... × ... = ...
$25 < x \leqslant 35$	50
$35 < x \leqslant 50$	15

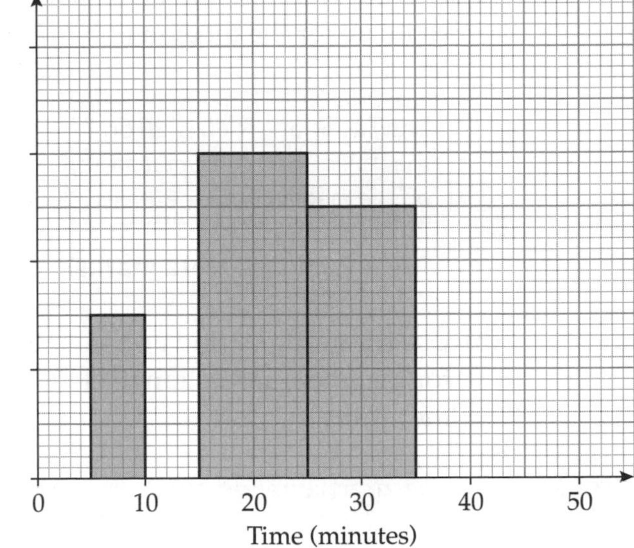

(a) Use the information in the histogram to complete the table.

Guided

Frequency density of $25 < x \leqslant 35$ class $= \dfrac{\text{frequency}}{\text{class width}}$

$= \dfrac{50}{10} = \text{.........}$

Use this information to put a scale on the y-axis.

Frequency = frequency density × class width

Use this formula to complete the table.

(2 marks)

(b) Use the information in the table to complete the histogram.

Use the frequency density to draw in the missing bars.

Guided

Frequency density of $10 < x \leqslant 15$ class $= \dfrac{20}{\text{......}} = \text{.........}$

Frequency density of $35 < x \leqslant 50$ class $= \dfrac{\text{......}}{\text{......}} = \text{.........}$

(2 marks)

 **2** The table gives information about the heights, in centimetres, of some students.

Height (h cm)	$145 < h \leqslant 155$	$155 < h \leqslant 160$	$160 < h \leqslant 175$	$175 < h \leqslant 190$
Frequency	30	20	75	45

(a) Use the table to draw a histogram.

(3 marks)

(b) Estimate the number of students who are taller than 170 cm.

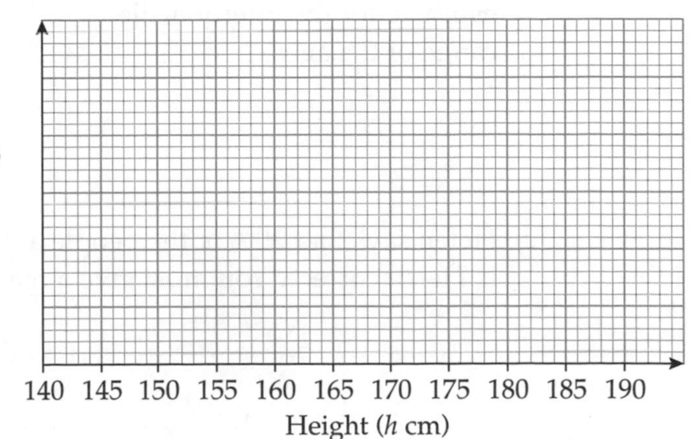

.... **(3 marks)**

Cumulative frequency

B **1** The cumulative frequency graph shows the marks scored by 60 students in a test.

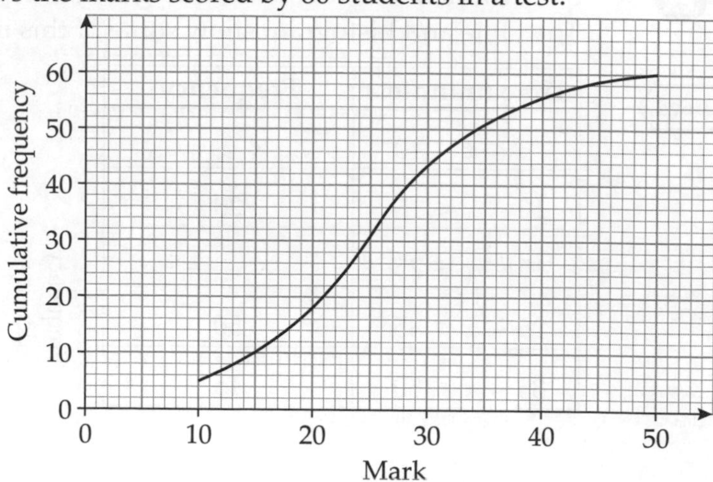

(a) Use the graph to find an estimate for the median test score.

> Draw a line at $\frac{1}{2} \times 60 = 30$ on the y-axis. Then you can read off the median from the x-axis.

.... **(2 marks)**

(b) Use the graph to find an estimate for the interquartile range of the scores.

Guided

Interquartile range = upper quartile − lower quartile

= .... − ....

= ....

> Take readings from the graph at $\frac{1}{4} \times 60 = 15$ for the lower quartile and $\frac{3}{4} \times 60 = 45$ for the upper quartile.

(2 marks)

B **2** The table gives information about the ages of the 80 people at a party.

Age (t years)	Cumulative frequency
$15 \leqslant t < 20$	10
$20 \leqslant t < 25$	42
$25 \leqslant t < 30$	63
$30 \leqslant t < 35$	75
$35 \leqslant t < 40$	80

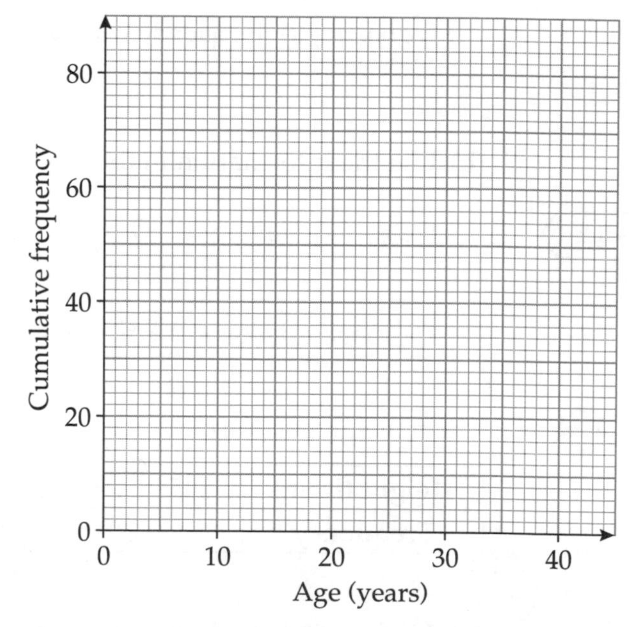

(a) On the grid, draw a cumulative frequency graph to show this information.

(2 marks)

(b) Use your graph to find an estimate for the interquartile range of the ages.

.... **(2 marks)**

(c) Helen says that 20% of the people at the party were older than 27
Is Helen correct? You must give a reason for your answer.

....

.... **(2 marks)**

Box plots

B 1 The box plot shows the heights, in cm, of some students in Year 11.

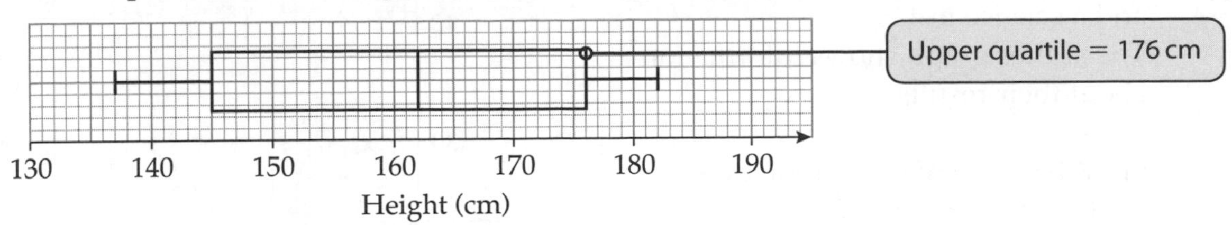

(a) Write down the median height.

Guided

Median height = ... cm **(1 mark)**

(b) Work out the interquartile range of the heights.

Guided

Interquartile range = upper quartile − lower quartile

= .... − ....

= .... **(2 marks)**

(c) There are 120 students in Year 11.
Work out the number of students taller than 176 cm.

The upper quartile is at 176 cm so 25% of the students will be taller than 176 cm.

Guided

Number of students taller than 176 cm = 25% of 120

= .... **(2 marks)**

B 2 The box plot gives information about the weights, in kilograms, of some cases.

Guided

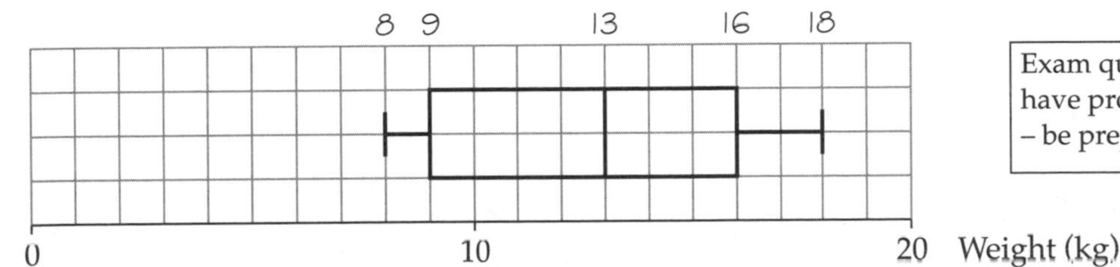

Exam questions similar to this have proved especially tricky – be prepared! ResultsPlus

Jean says that exactly 50% of the cases have weights between 13 kg and 16 kg. Is Jean correct?
You must give a reason.

13 kg = median (......... %)

16 kg = upper quartile (......... %)

Jean is ....

....

(2 marks)

B 3 The table shows information about the marks attained by students in a test.

Lowest mark	8
Lower quartile	27
Median mark	35
Upper quartile	46
Range	62

Use the lowest mark and the range to work out the highest mark.

Use this information to draw a box plot.

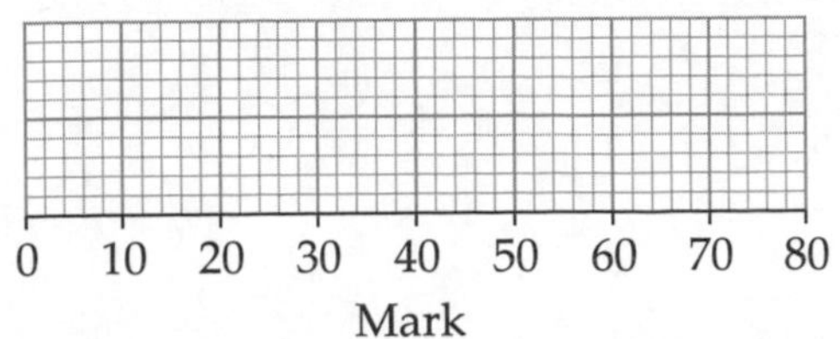

(3 marks)

Scatter graphs

D **1** Some students took both an English test and a science test.
The scatter graph shows information about their results.

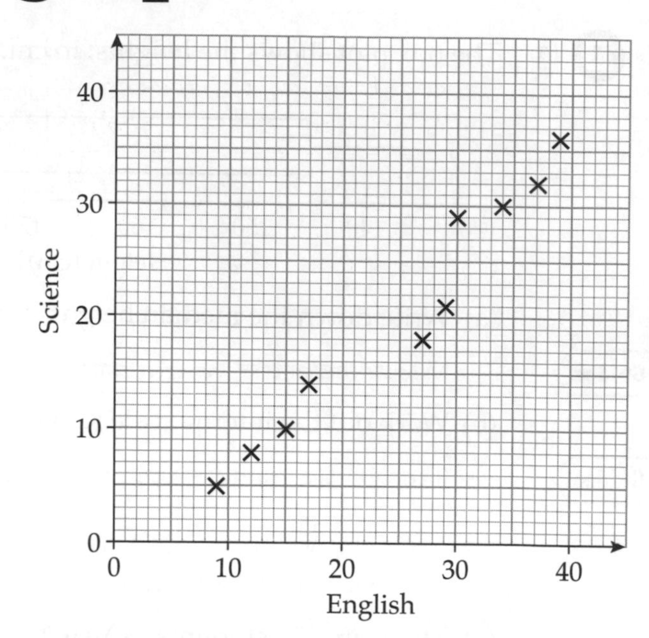

(a) What type of correlation does this scatter graph show?

.... **(1 mark)**

(b) Draw a line of best fit on the scatter graph.

(1 mark)

(c) Use your line of best fit to estimate the science mark for a student with an English mark of 25

Draw a vertical line from a mark of 25 for English to your line of best fit. Then draw a horizontal line across to find the science mark.

.... **(2 marks)**

D **2** The scatter graph shows some information about the ages and values of 10 cars.
The cars are the same make and type.

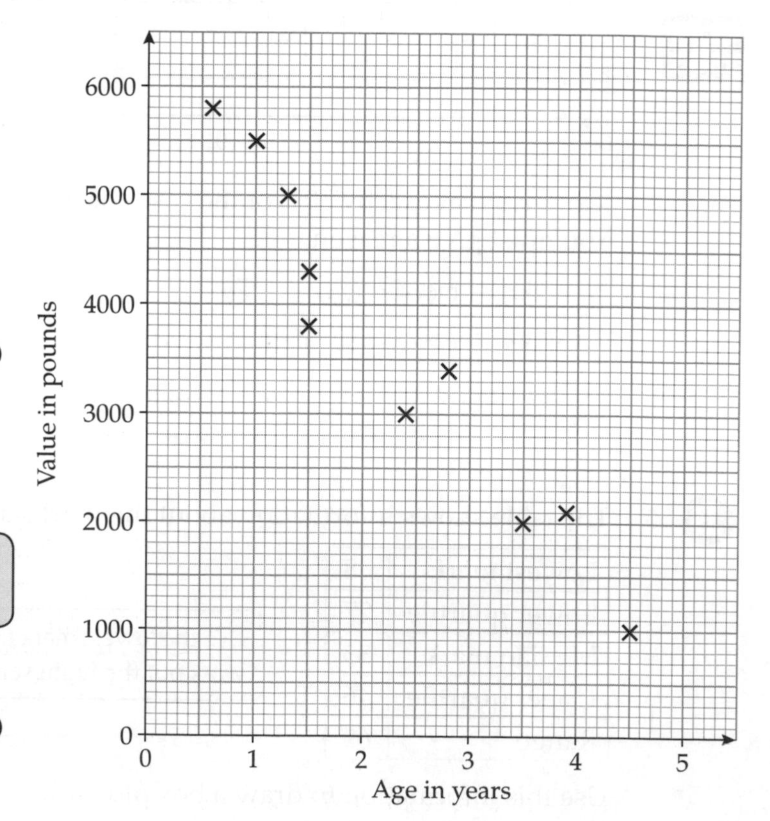

(a) What type of correlation does this scatter graph show?

.... **(1 mark)**

(b) Another car of the same make and type is 2 years old.
Estimate the value of this car.

Draw a line of best fit on the graph to help you estimate the value.

£....................... **(2 marks)**

(c) Another car of the same make and type has a value of £2800
Estimate the age of this car.

.... **(2 marks)**

Probability

D 1 Four teams, City, Rovers, Town and United play a competition to win a cup.
Only one team can win the cup.
The table below shows the probabilities of City or Rovers or Town winning the cup.

City	Rovers	Town	United
0.38	0.27	0.15	x

Work out the value of x.

Guided

Probability(United) = 1 − Probability(not United)

= 1 − (0.38 + 0.27 + 0.15)

= 1 − ....

= .... **(2 marks)**

2 The probability that a biased dice will land on a 2 is 0.3
Sue is going to roll the dice 60 times.
Work out an estimate for the number of times the dice will land on a 2

Guided

Estimate for number of 2s = ... × 60

= .... **(2 marks)**

3 A bag contains counters which are red or green or yellow or blue.
The table shows each of the probabilities that a counter taken at random from the bag will be red or green or blue.

Colour	Red	Green	Yellow	Blue
Probability	0.15	0.2		0.3

A counter is to be taken at random from the bag.

(a) Work out the probability that the counter will be yellow.

.... **(2 marks)**

The bag contains 200 counters.

(b) Work out the number of red counters in the bag.

Make sure that you select the correct probability from the table.

.... **(2 marks)**

4 Marco has a 4-sided spinner.
The sides of the spinner are numbered 1, 2, 3 and 4
The spinner is biased.
The table shows the probability that the spinner will land on each of the numbers 1, 2 and 4

Number	1	2	3	4
Probability	0.2	0.35		0.18

Marco spins the spinner 400 times.
Work out an estimate for the number of times the spinner lands on 3

.... **(3 marks)**

Tree diagrams

A **1** Jill is going to play one game of tennis and one game of badminton.
The probability that she will win the game of tennis is $\frac{1}{4}$
The probability that she will win the game of badminton is $\frac{2}{5}$

(a) Complete the probability tree diagram.

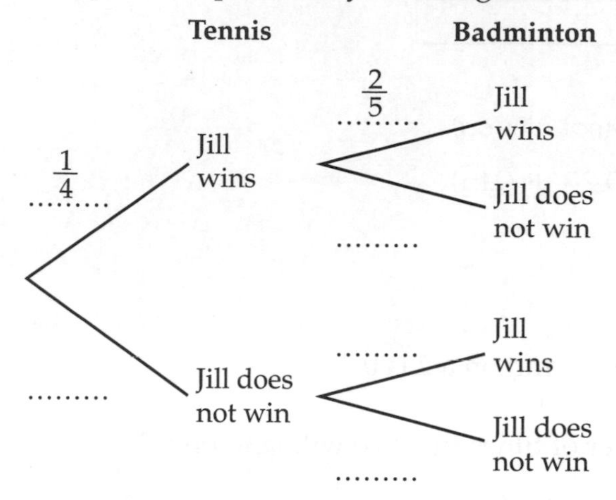

Remember that at each branch the probabilities add up to 1.

(2 marks)

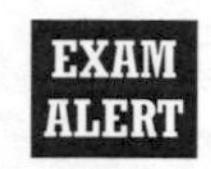
EXAM ALERT

Guided

(b) Work out the probability that Jill wins at least one game.

P(Jill wins at least one game) = 1 − P(lose tennis, lose badminton)

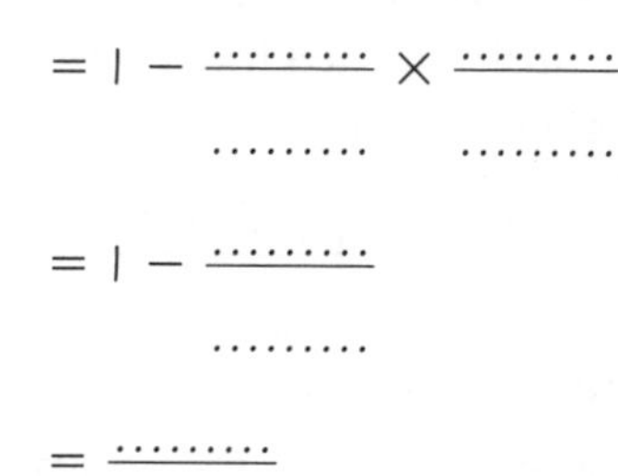

$= 1 - \frac{\dots\dots\dots}{\dots\dots\dots} \times \frac{\dots\dots\dots}{\dots\dots\dots}$

$= 1 - \frac{\dots\dots\dots}{\dots\dots\dots}$

$= \frac{\dots\dots\dots}{\dots\dots\dots}$

Exam questions similar to this have proved especially tricky – be prepared! ResultsPlus

(3 marks)

2 There are 4 girls and 6 boys in a music group.
The teacher selects two of the students at random.
Work out the probability that he selects two boys.

This is a non-replacement question as the teacher cannot select the same boy twice.

.... **(3 marks)**

A* **3** There are 4 orange sweets, 2 red sweets and 3 green sweets in a bag.
Martha takes a sweet at random.
She eats the sweet.
She then takes and eats another sweet at random.
Work out the probability that the sweets are **different** colours.

.... **(4 marks)**

Problem-solving practice

D **1** All students in class 9A study either French or German or Spanish.
21 of the 40 students are boys.
7 of the boys study Spanish.
9 girls study French.
8 of the 10 students who study German are boys.
Work out the number of students who study Spanish.

Draw a two-way table. Label the columns 'French', 'German' and 'Spanish'. Label the rows 'Boys' and 'Girls'. You will also need a column and row for the totals.

.... students study Spanish. **(4 marks)**

D ***2** Tom picks 10 apples from each of two apple trees of the same type.
He weighs each of the apples.

	Weight in grams									
Tree A	45	60	38	65	60	55	52	54	42	47
Tree B	41	46	53	57	62	63	61	56	58	43

Compare fully the weights of the apples from the two trees.

For each tree, find (1) the mean **or** median weight of the apples and (2) the range of the weights of the apples. Write a sentence comparing the mean **or** the median weights and interpret these results. **And** write a sentence comparing the range of the weights and interpret these results.

The question has a * next to it, so make sure that you show all your working and write your answer clearly with the correct units.

(6 marks)

C **3** There are 12 boys and 10 girls in a class.
All the students take a French test.
The mean mark for the girls is 58
The mean mark for the boys is 64
Work out the mean mark for the class as a whole.

.... **(3 marks)**

Problem-solving practice

4 The box plot gives information about the weights, in kilograms, of some bags.
Katie says that 25% of all the bags have a weight greater than 18.2 kg.
Is Katie correct? You must give a reason.

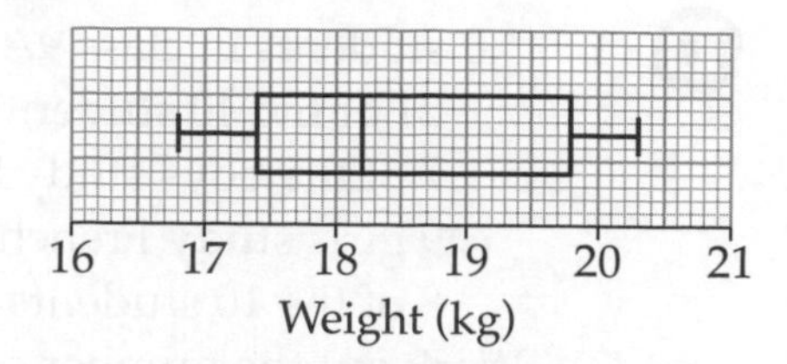

The upper quartile will give the minimum weight for the top 25% of bags.

Guided

Upper quartile = ... kg

Katie is .... because the upper quartile is ....

An alternative method of working would be to look at the value of the median and use this in your answer.

(2 marks)

5 The histogram gives information about the weights of 220 letters.

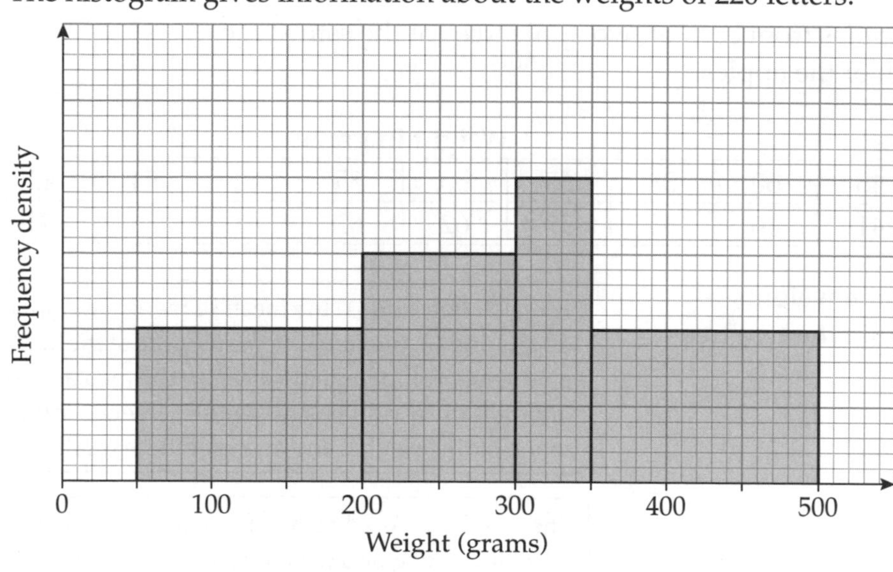

First work out the number of letters represented by each cm square: count the number of squares and divide 220 by this number. Next count the number of squares that represent letters with a weight greater than 250 g. Multiply this total by the number of letters represented by each square.

Work out an estimate for the number of letters with a weight greater than 250 grams.

.... **(3 marks)**

6 There are 5 blue pens and 3 red pens in a box.
Liz takes **two** of the pens at random.
Work out the probability that both pens will be the same colour.

You could draw a tree diagram to help answer this question. Since it is a **non-replacement** probability question, the denominators of the fractions on the second sets of branches will be 1 less than those on the first set.

Find the probability of Liz taking two red pens and the probability of Liz taking two blue pens. Add these fractions together for your final answer.

.... **(4 marks)**

Formulae page

Volume of a prism = area of cross section × length

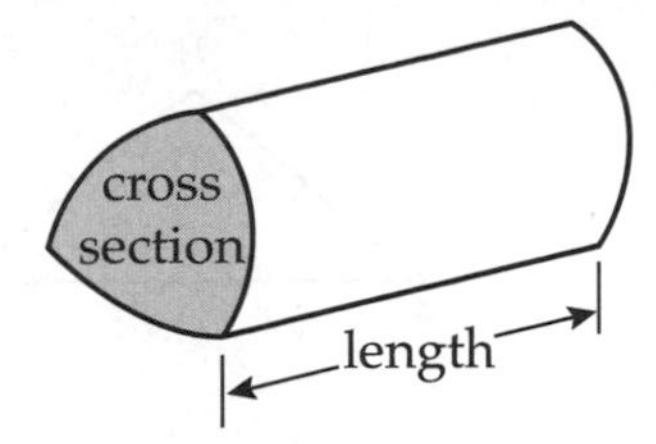

Area of trapezium = $\frac{1}{2}(a + b)h$

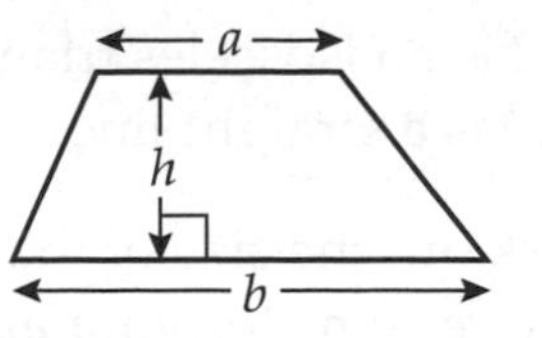

Volume of sphere = $\frac{4}{3}\pi r^3$

Surface area of sphere = $4\pi r^2$

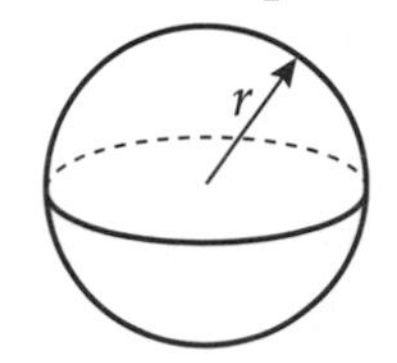

Volume of cone = $\frac{1}{3}\pi r^2 h$

Curved surface area of cone = πrl

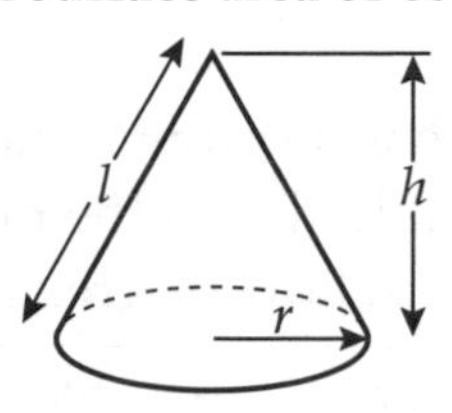

In any triangle ABC

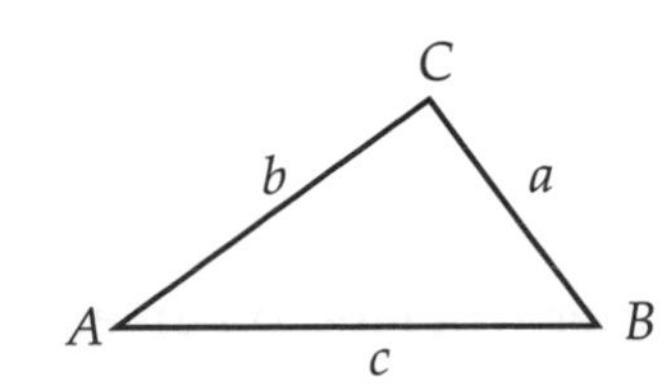

Sine Rule $\frac{a}{\sin A} = \frac{b}{\sin B} = \frac{c}{\sin C}$

Cosine Rule $a^2 = b^2 + c^2 - 2bc \cos A$

Area of triangle = $\frac{1}{2}ab \sin C$

The Quadratic Equation

The solutions of $ax^2 + bx + c = 0$

where $a \neq 0$, are given by

$$x = \frac{-b \pm \sqrt{(b^2 - 4ac)}}{2a}$$

Paper 1

Practice exam paper

Higher Tier
Time: 1 hour 45 minutes
Calculators must not be used

1 ABC is an isosceles triangle.
ACD is a straight line.

Work out the size of the angle marked x.
Give reasons for your answer.

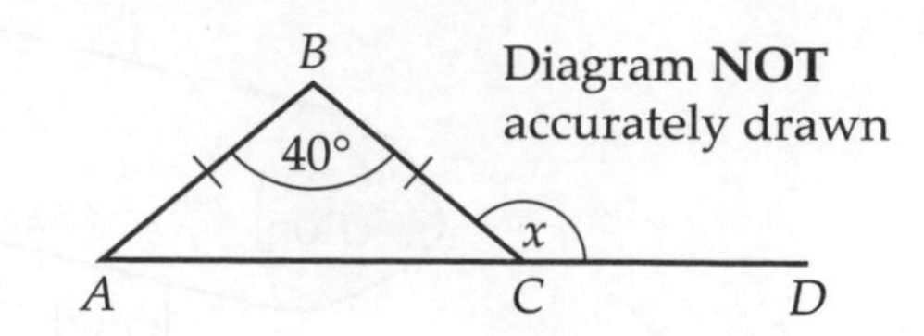

(3 marks)

2 Cara has a spinner that is biased.
The table shows the probabilities that the spinner will land on red or yellow or green.

Colour	red	blue	green	yellow
Probability	0.24		0.4	0.2

(a) Work out the probability that the spinner will land on blue.

.... **(2 marks)**

Cara is going to spin the spinner 80 times.

(b) Work out an estimate for the number of times the spinner will land on green.

.... **(2 marks)**

3 On the grid, enlarge the shaded shape with scale factor 2

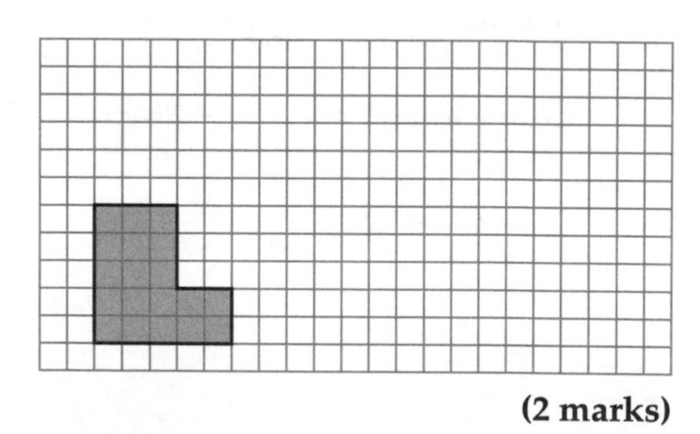

(2 marks)

4 (a) Simplify $5(2x - 3y) + 2(x - 4y)$

.... **(2 marks)**

(b) Factorise fully $12y^2 + 8y$

.... **(2 marks)**

(c) Expand and simplify $(x + 6)(x - 8)$

.... **(2 marks)**

5 Here are the first five terms of an arithmetic sequence.

7 13 19 25 31

(a) Write down an expression, in terms of n, for the nth term of this sequence.

.... **(2 marks)**

Janice says that 104 is a term in this sequence.

(b) Is Janice correct? You must give a reason for your answer.

....

(1 mark)

6 On the grid, draw the graph of $y = 3x - 4$ for values of x from -1 to 3

(3 marks)

7 The table gives information about the costs of sending a letter by first-class post and by second-class post.

First class	Second class
46p	36p

In one week, a company sent 400 letters.
The ratio of the number of letters sent first class to the number of letters sent second class was 5 : 3
Work out the total cost of sending all the letters.

£.................... **(4 marks)**

8 Berwyn wants to find out how many magazines people buy.
He is going to carry out a survey using a questionnaire.

(a) Design a suitable question Berwyn could use for his questionnaire.

(2 marks)

(b) Berwyn stands outside a newspaper shop.
He gives a questionnaire to all the women leaving the shop.
His sample is biased. Give **two** reasons why.

1 ....

....

2 ....

....

(2 marks)

9 Abby is buying food for a barbecue.
She is going to make burger buns. She needs a bread roll and a burger for each burger bun.
There are 36 bread rolls in a pack.
There are 24 burgers in a pack.
Abby wants to make at least 200 burger buns.
Abby wants to buy exactly the same number of bread rolls and burgers.

(a) What are the smallest numbers of packs of bread rolls and packs of burgers she can buy?

.... packs of bread rolls

.... packs of burgers

(b) How many burger buns will she make?

.... **(5 marks)**

10 (a) Solve $5(y - 2) = 8$

.... **(2 marks)**

(b) Solve $6x - 4 = 4x + 13$

.... **(3 marks)**

(c) Factorise $6x^2 - 13x - 5$

.... **(2 marks)**

11 A path is measured to the nearest metre. The length is given as 18 m.

(a) What is the minimum possible length of the path?

.... m

(b) What is the maximum possible length of the path?

.... m **(2 marks)**

12 Change $3\,m^2$ to cm^2.

.... cm^2 **(2 marks)**

13 Solve the simultaneous equations

$3x + 5y = -1$

$4x - 2y = 16$

$x =$ ...., $y =$ .... **(4 marks)**

14 The diagram shows two towns, Alton and Beescroft.

× B

A×

Scale: 1 cm represents 10 km

A new airport is to be built. The airport must be

less than 70 km from Beescroft **and**

closer to Alton than to Beescroft.

On the diagram, shade the region where the airport could be built.

(3 marks)

15 Work out $6\frac{2}{5} \div 5\frac{1}{3}$

.... **(3 marks)**

16 Mr Smith bought a television. The usual price of the television was £850
The television was in a sale. In the sale, all prices were reduced by 20%.
Mr Smith paid £250 when he got the television.
He paid the rest of the cost of the television in 10 equal monthly payments.
Work out the amount of each monthly payment.

£.................... **(5 marks)**

17 (a) Find the value of 9^0 (b) Find the value of $64^{\frac{1}{2}}$ (c) Find the value of 2^{-4}

.... **(1 mark)** .... **(1 mark)** .... **(1 mark)**

18 The cumulative frequency graph shows the amounts spent by 80 customers at a supermarket.

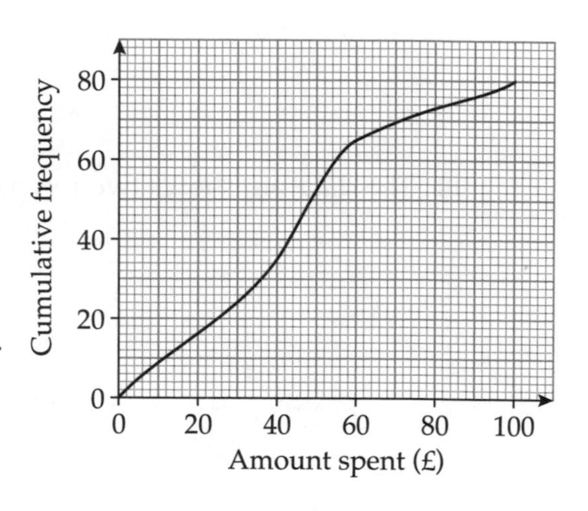

(a) Work out an estimate for the median amount spent.

.... **(2 marks)**

(b) Work out an estimate for the interquartile range.

.... **(2 marks)**

(c) Work out an estimate for the number of people who spent **more** than £75

.... **(2 marks)**

(d) Explain why the interquartile range may be a better measure of spread than the range.

....

....

(1 mark)

***19** A, B, C and D are points on the circumference of a circle, centre O.
Angle $AOC = 124°$.
Work out the value of x. Give reasons for each stage of your working.

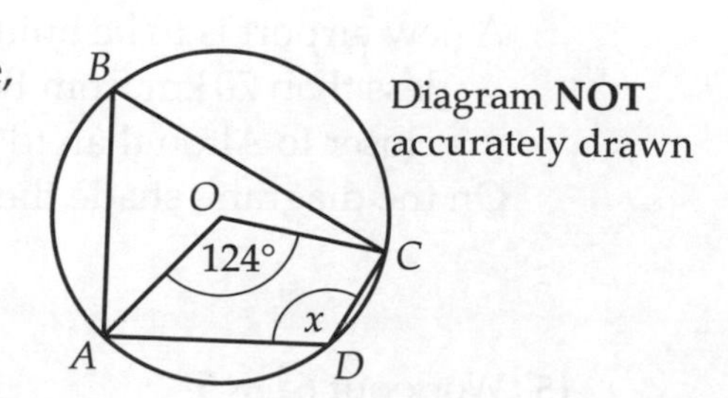

(4 marks)

20 A is the point (0, 3).
B is the point (8, 6).
Find an equation of the straight line perpendicular to AB passing through A.

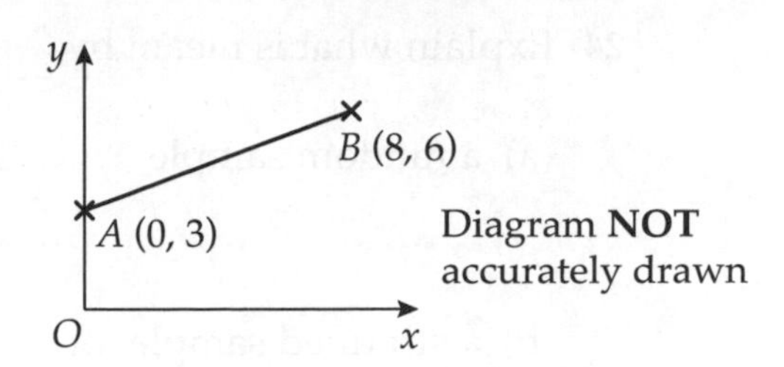

.... **(3 marks)**

***21** The diagram shows a six-sided shape.
All the corners are right angles.
All the measurements are given in centimetres.
Show that the area of this shape can be written as
$5x^2 + 4x - 9$

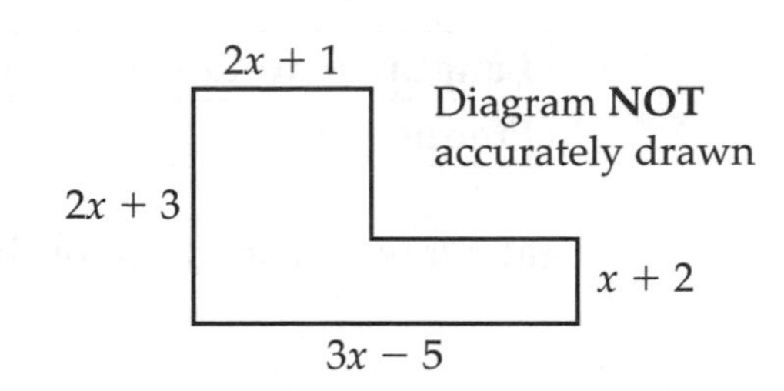

(4 marks)

***22** Express the recurring decimal $0.4\dot{5}\dot{2}$ as a fraction in its simplest form.

(3 marks)

23 $OABC$ is a trapezium.
OA is parallel to CB. $CB = 3OA$.
$\overrightarrow{OA} = \mathbf{a}$. $\overrightarrow{OC} = \mathbf{c}$.
P is the point on CB such that $CP : PB = 2 : 1$

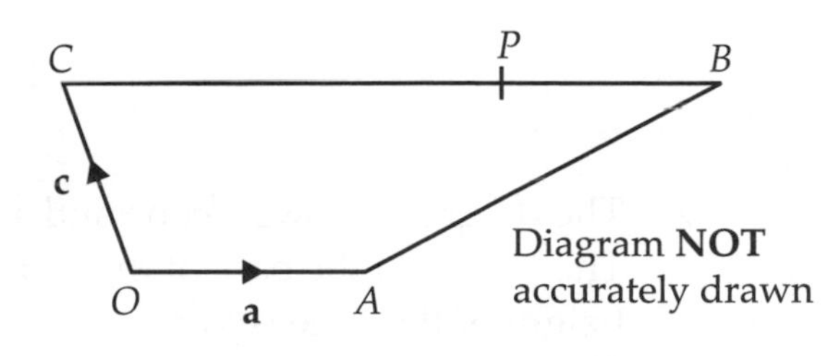

(a) Express $\overrightarrow{OB}$ in terms of **a** and **c**.
Give your answer in its simplest form.

.... **(2 marks)**

(b) Express $\overrightarrow{AP}$ in terms of **a** and **c**. Give your answer in its simplest form.

.... **(3 marks)**

24 Explain what is meant by

(a) a random sample ....

....

(b) a stratified sample..

....

(2 marks)

25 A customer helpline receives 110 calls one day.
The table gives information about the lengths, in minutes, of these telephone calls.

Length (m minutes)	$0 < m \leqslant 5$	$5 < m \leqslant 15$	$15 < m \leqslant 30$	$30 < m \leqslant 40$
Frequency	6	25	60	19

(a) Draw a histogram for this information.

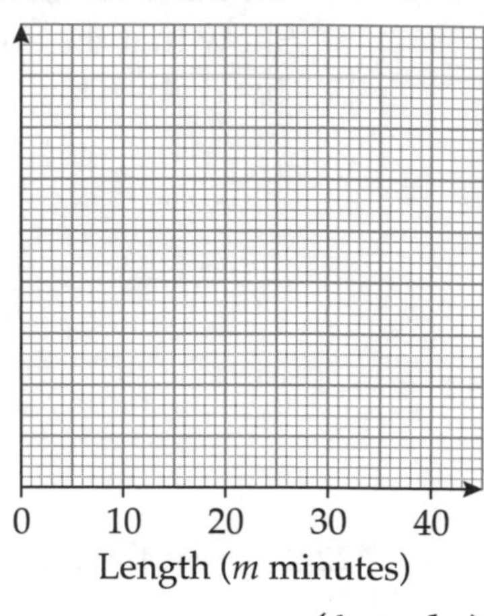

(4 marks)

(b) Use your histogram to work out an estimate for the number of calls that lasted more than 20 minutes.

.... **(2 marks)**

26 The diagram shows a cone and a sphere.
The radius of the base of the cone is $3x$ cm and the height of the cone is h cm.
The radius of the sphere is $2x$ cm.
The volume of the cone is equal to the volume of the sphere.
Express h in terms of x.
Give your answer in its simplest form.

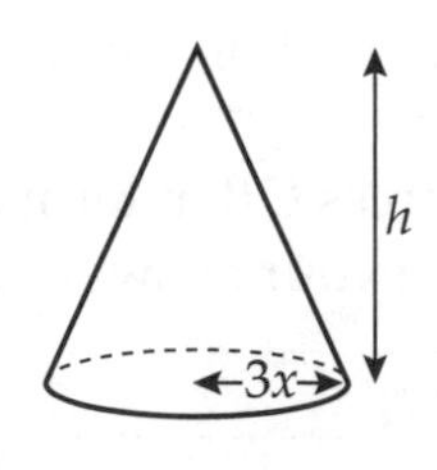

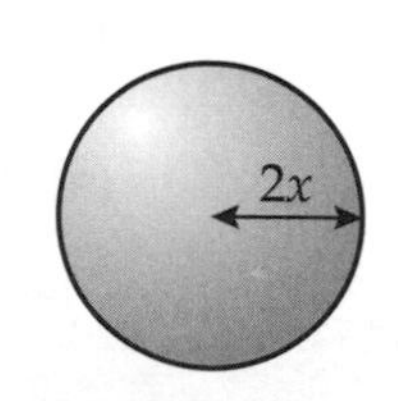

Diagram **NOT** accurately drawn

$h =$ .... **(3 marks)**

Paper 2

Practice exam paper

Higher Tier
Time: 1 hour 45 minutes
Calculators may be used

1 Here is a list of ingredients needed to make 8 pancakes.
Sally wants to make 20 pancakes.
Work out the amount of each ingredient she will need.

Ingredients for 8 pancakes
120 g plain flour
2 eggs
200 m*l* milk
50 g butter

Plain flour .... g Eggs ....

Milk .... m*l* Butter .... g **(3 marks)**

2 The cost of 5 kg of apples is £4.60
The cost of 3 kg of apples and 4 kg of pears is £6.16
Work out the cost of 1 kg of pears.

.... **(4 marks)**

3 (a) Use your calculator to work out the value of $\sqrt{\dfrac{45.2 \times 17}{13.4 - 6.9}}$

Write down all the figures on your calculator display.

.... **(3 marks)**

(b) Write down your answer to part (a) correct to 3 significant figures.

.... **(1 mark)**

4 40 students took a test. The test is marked out of 50
The table gives information about the students' marks.

Mark	0–10	11–20	21–30	31–40	41–50
Frequency	6	8	9	12	5

On the grid, draw a frequency polygon to show this information.

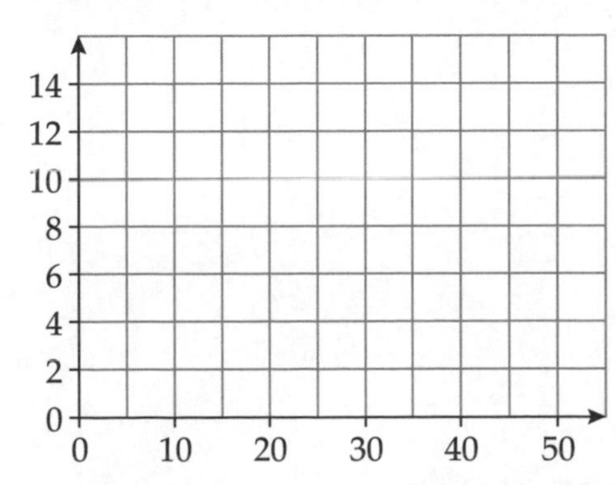

(3 marks)

5 (a) Simplify $x^8 \times x^3$

.... **(1 mark)**

(b) Simplify $y^9 \div y^3$

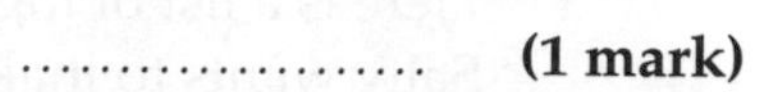

.... **(1 mark)**

(c) Simplify $(m^7)^2$

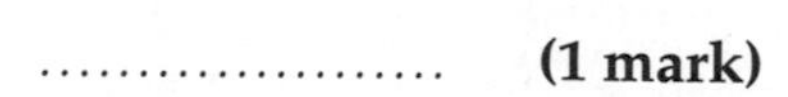

.... **(1 mark)**

6 The diagram shows a sketch of Mr Warder's garden.
Mr Warder wants to spread grass seed all over his garden.
One box of grass seed is enough to cover $10\,m^2$.
One box of grass seed costs £6.48
Mr Warder wants to spend as little as possible.
How much will he have to pay for grass seed?
You must show all your working.

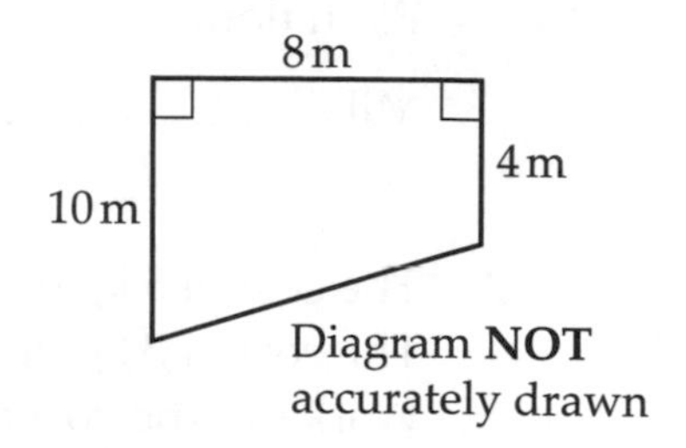

.... **(5 marks)**

7 (a) Reflect shape **Q** in the line $y = -1$
Label the new shape **R**.

(2 marks)

(b) Describe fully the single transformation that maps shape **P** onto shape **Q**.

....

....

....

(3 marks)

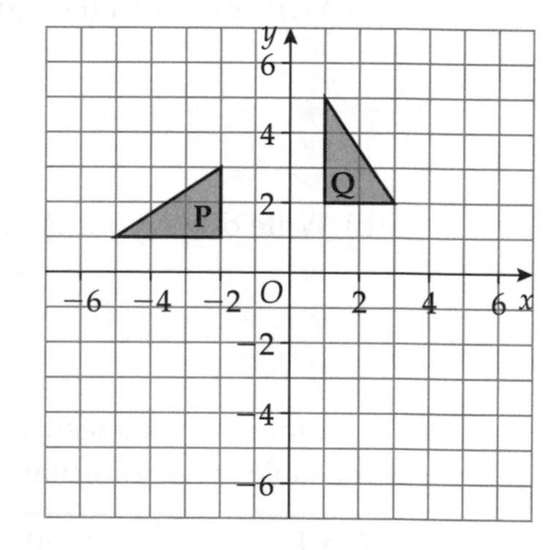

8 On the grid, draw the graph of $y = 2x^2 - 8$ for values of x from −3 to 3

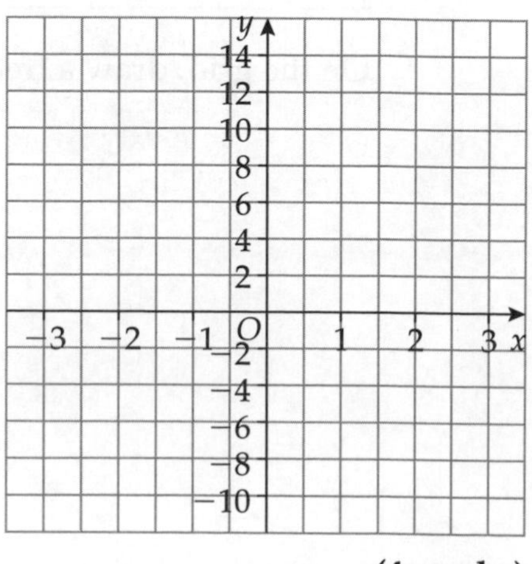

(4 marks)

9 (a) Write 0.00902 in standard form.

.... **(1 mark)**

(b) Write 6.08×10^5 as an ordinary number.

.... **(1 mark)**

(c) Work out $4.5 \times 10^6 \times 2.6 \times 10^{-2}$

.... **(1 mark)**

***10** Liam needs to buy a ladder. He wants to use the ladder to reach a window.
The ladder will have to reach a point that is 5.5 metres above the ground.
He will place the ladder on horizontal ground. It will rest against a vertical wall.
The ladder cannot be placed closer than 1.4 m to the wall.
Liam finds a 6 metre ladder on sale.
Is this ladder long enough to reach the window? You must show all your working.

(4 marks)

11 The diagram shows the floor of a room.
The floor is in the shape of a semicircle of diameter 16 m.
The floor is to be varnished.
Two coats of varnish are to be put on the floor.
One tin of varnish will cover an area of $20\ m^2$.
How many tins of varnish will be needed?

Diagram **NOT** accurately drawn

.... **(4 marks)**

12 The equation $x^3 - 6x = 140$ has a solution between 5 and 6
Use a trial and improvement method to find this solution.
Give your answer correct to 1 decimal place.
You must show **ALL** your working.

$x =$ .... **(4 marks)**

13 The table shows the times taken by a group of 40 students to complete a puzzle.

Time (t seconds)	$30 < t \leqslant 40$	$40 < t \leqslant 50$	$50 < t \leqslant 60$	$60 < t \leqslant 70$	$70 < t \leqslant 80$
Frequency	7	15	10	5	3

(a) Find the class interval that contains the median.

.... **(2 marks)**

(b) Calculate an estimate for the mean time taken by the students.

.... **(4 marks)**

One further student completed the puzzle in a time of 90 seconds.
Raki says, 'The class interval that contains the median will now change.'

(c) Is Raki correct? Explain your answer.

....

....

(1 mark)

14 Calculate the value of x.
Give your answer correct to 1 decimal place.

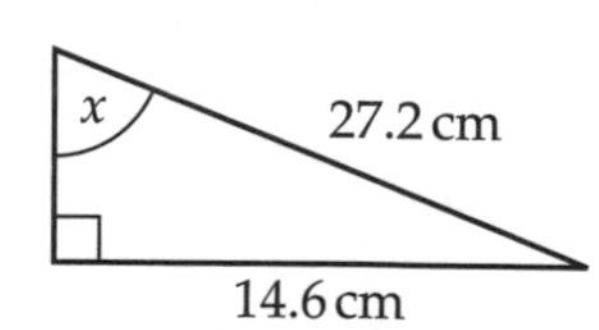

Diagram **NOT** accurately drawn

$x =$ ....° **(3 marks)**

15 In a sale, normal prices are reduced by 35%.
The sale price of a computer is £312
Work out the normal price of the computer.

£.................... **(3 marks)**

16 Make a the subject of the formula

$$4(a - 2) = t(1 - 2a)$$

$a =$ .... **(4 marks)**

17 Dom invested £6000 for 3 years in a savings account.
He was paid 2.7% per annum compound interest.
How much did Dom have in his savings account after 3 years?

£.................... **(3 marks)**

***18** Martin recorded the times that 15 boys and 15 girls took to complete a puzzle.

		Boys				Girls			
				25	6				
	8	7	1	26	1	4	5	7	
	7	5	3	27	4	5	7	8	9
6	6	4	4	28	0	0	2	3	
	1	1	0	29	6				
		7	5	30					

Key: 5|26 means 26.5 seconds   Key: 26|5 means 26.5 seconds

Compare the times taken by the boys with the times taken by the girls.

(6 marks)

19 Solve $3x^2 - 5x - 1 = 0$

Give your solutions correct to 3 significant figures.

.... **(3 marks)**

20 Correct to 2 significant figures, the area of a rectangle is 340 cm^2.

Correct to 2 significant figures, the length of the rectangle is 17 cm.

Calculate the upper bound for the width of the rectangle.

.... **(3 marks)**

21 d is inversely proportional to the square of c. When $c = 0.2$, $d = 375$

Find the value of d when $c = 5$

$d =$ .... **(3 marks)**

22 $AB = 23$ m. $AC = 75$ m. Angle $ABC = 104°$.

Work out the area of the triangle.

Give your answer correct to 3 significant figures.

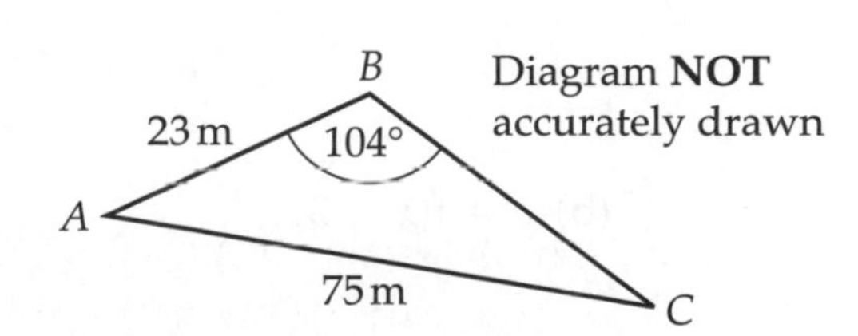

.... m^2 **(5 marks)**

23 The table shows the numbers of students in three year groups.

Year group	Number of students
Year 11	246
Year 12	180
Year 13	174

Andy carries out a survey of the students in Years 11, 12 and 13.
He uses a sample stratified by year group.
There are 30 students from Year 12 in his sample.
Work out the number of students from Year 11 in his sample.

.... **(2 marks)**

24 Write as a single fraction in its simplest form.

$$\frac{x}{2x-1} - \frac{8x-3}{4x^2-1}$$

.... **(4 marks)**

25 A bag contains 5 red, 3 yellow and 4 blue beads.
Josip takes three beads at random from the bag and puts them in a box.
Work out the probability that at least 2 beads in the box are the same colour.

.... **(5 marks)**

26 The diagram shows part of the curve with equation $y = \mathrm{f}(x)$.
The coordinates of the maximum point of this curve are $(-2, 5)$.
Write down the coordinates of the maximum point of the curve with equation

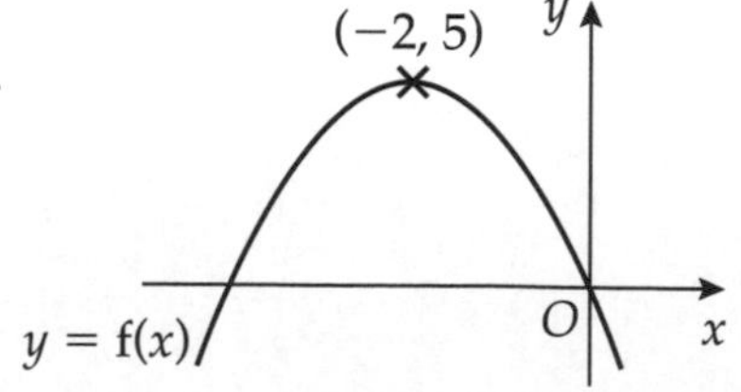

(a) $y = \mathrm{f}(x) - 2$

(..........., ...) **(1 mark)**

(b) $y = \mathrm{f}(x - 3)$

(..........., ...) **(1 mark)**

(c) $y = \frac{1}{2}\mathrm{f}(x)$

(..........., ...) **(1 mark)**

Answers

The number given to each topic refers to its page number.

NUMBER

1. Factors and primes

1. (a) (i) $2 \times 2 \times 3 \times 5$ (ii) $2 \times 3 \times 5 \times 5$
   (b) 30 (c) 300
2. (a) $2 \times 2 \times 2 \times 3 \times 3$ (b) 24 (c) 360

2. Indices 1

1. (a) 7^{13} (b) 7^6 (c) 7^7 (d) 7^{20}
2. (a) 5^{12} (b) 5^6 (c) 5^6
3. 6 4. 1 5. 3^{12} 6. 2^{17}

3. Fractions

1. $5\frac{7}{15}$
2. (a) $4\frac{10}{21}$ (b) $11\frac{1}{2}$
3. $3\frac{11}{15}$
4. (a) $7\frac{1}{2}$ (b) $4\frac{4}{11}$
5. (a) $5\frac{13}{30}$ (b) $4\frac{2}{3}$

4. Decimals

1. (a) 19.43 (b) 1.943 (c) 290
2. (a) 106 680 (b) 1.0668 (c) 8400
3. (a) 30 240 (b) 0.3024 (c) 4800

5. Recurring decimals

1. $\frac{5}{33}$ 2. $\frac{8}{9}$
3. $2\frac{139}{333}$ 4. $\frac{43}{90}$
5. $x = 0.827\,27\ldots$; $100x = 82.727\ldots$; $99x = 81.9$; $x = \frac{819}{990} = \frac{91}{110}$

6. Rounding and estimation

1. (a) 8000 (b) 10 (c) 1500
2. 100
3. 48 000
4. 1600
5. 2000
6. (a) 20, overestimate (b) 5, underestimate
7. 300

7. Upper and lower bounds

1. 63.8625 cm^2
2. 66.1 cm
3. (a) 4.8636, 4.9094 (b) 4.9 s

8. Fractions and percentages

1. £100
2. (a) 62.5% (b) $\frac{3}{8}$
3. £640
4. £10 400
5. 60 students

9. Percentage change

1. £504 2. £54.60
3. 2.5% 4. 35p
5. £92.95

10. Reverse percentages and compound interest

1. £9396.11 2. £3547.90 3. £620
4. Her weekly pay last year would have been £400 if she had a 15% pay rise **or** if her pay last year was £391 then with a 15% increase it would be £449.65 this year
5. $n = 6$

11. Ratio

1. 7 : 3 2. No – the ratio is 20 : 12 = 5 : 3
3. £90 4. £40
5. 720 km

12. Proportion

1. £9.79 2. £17.40
3. £0.95 or 95p 4. 195 minutes or $3\frac{1}{4}$ hours
5. 6 days 6. 20 hours

13. Indices 2

1. (a) $\frac{1}{16}$ (b) 7
2. (a) 3 (b) $\frac{1}{9}$ (c) $\frac{1}{64}$ (d) 1
3. (a) 4 (b) $\frac{64}{27}$
4. (a) $\frac{1}{7}$ (b) 16 (c) $\frac{8}{27}$
5. 1

14. Standard form

1. (a) 6.7×10^4 (b) 0.000 02 (c) 7.6×10^6
2. (a) 5.4×10^{-1} (b) 7 000 000
3. 32×10^6, 3×10^8, 0.031×10^{10}, 3400×10^5
4. 4.5×10^{11} 5. 5.2×10^7
6. 2.1×10^9 7. 4.53×10^{27}

15. Calculator skills

1. 25.390 804 6 2. 1.451 013 451
3. 2.526 844 327 4. 2.65×10^3
5. 1.6×10^{-2}

16. Surds

1. (a) $4\sqrt{3}$ (b) $10\sqrt{3}$
2. $5\sqrt{2}$ 3. $7 - 3\sqrt{3}$
4. $9 + 5\sqrt{5}$ 5. $11 - 6\sqrt{2}$
6. $-5 + 4\sqrt{3}$

17–18. Problem-solving practice

1. Cheap Tickets (£37.63, compared with £37.80 at Tickets R-US)
2. £10 400
3. 4 boxes
4. Simple Bank pays more interest (£810, compared with only £788.20 from Compound Bank).
5. 9.10 miles per litre

ALGEBRA

19. Algebraic expressions

1. (a) m^{12} (b) p^8 (c) t^{20}
2. (a) g^7 (b) k^8 (c) y^{21}
3. (a) x^3 (b) y^8 (c) z^6
4. (a) $15e^{11}f^7$ (b) $4x^5y^2$ (c) $16m^{20}p^4$
5. (a) $24c^6d^{10}$ (b) $5a^6c$ (c) $125b^9d^6$
6. (a) $4x^6$ (b) $\frac{8bc^4}{5}$

20. Arithmetic sequences

1. $4n - 3$ 2. $-5n + 22$
3. (a) $4n - 1$
   (b) 72 is an even number, all numbers in sequence are odd numbers
4. (a) 11, 19, 27
   (b) 5th term is 43, 6th term is 51 therefore 45 is not a term
5. 149

21. Expanding brackets

1. (a) $10x + 7y$
   (b) $2m^4 - 5m$
2. $5x - y$
3. (a) $18 - 24d$ (b) $3p^3 - p^2$
4. (a) $x^2 + 4x - 21$ (b) $y^2 + 12y + 36$
5. (a) $p^2 - 9p + 20$ (b) $t^2 - 16t + 64$
6. $6x^2 + 2x - 20$
7. (a) $10x^2 - 43x + 28$ (b) $9y^2 - 24y + 16$
   (c) $14x^2 - 15xy - 9y^2$

22. Factorising

1. (a) $6(x + 3)$ (b) $y(y - 9)$
2. (a) $5(m + 4)$ (b) $v(3 - v)$
3. $4m(2p - 3m)$
4. (a) $7b(2a - 3c)$ (b) $8xy(2x - 5y)$
5. (a) $(x + 5)(x + 4)$ (b) $(x + 5)(x - 5)$
6. (a) $(y - 6)(y + 2)$ (b) $(y - 1)(y + 1)$
7. (a) $(3x + 7)(x + 1)$ (b) $(5x + 3)(5x - 3)$
8. (a) $(5x - 3)(x + 4)$ (b) $(2x + 7)(2x - 7)$
9. (a) $(2x + 3y)(x + y)$ (b) $(4x - 3y)(2x + y)$

23. Linear equations 1

1. $x = 7$
2. (a) $x = 4$ (b) $x = \frac{1}{2}$
3. (a) $x = \frac{13}{4}$ (b) $x = -17$ 4. 140° 5. 7.5 cm

24. Linear equations 2

1. $x = -\frac{13}{2}$ 2. $x = -3$
3. $x = 9$ 4. $x = \frac{16}{7}$
5. $x = \frac{5}{11}$

25. Straight-line graphs

1.

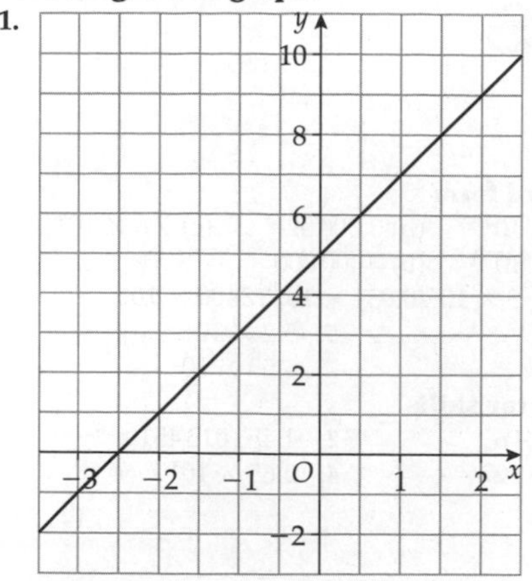

2. (a) −2 (b) (0, 3)
3. (a) $y = 2x + 3$ (b) $y = 4 - x$ (c) $y = \frac{1}{2}x - 3$
4. $y = \frac{1}{2}x + 3$

26. Parallel and perpendicular

1. $(3, 5\frac{1}{2})$ 2. (2, 3) 3. e.g. $y = 4x + 1$
4. e.g. $y = -\frac{1}{2}x + 3$ 5. $y = 3x + 4$ 6. $y = -\frac{1}{2}x + 3$

27. 3-D coordinates

1. (a) B (b) (3, 2, 1) 2. (5, 2, 0) 3. $(6, -1, \frac{1}{2})$
4. (a) (5, 1, 2) (b) $(2\frac{1}{2}, \frac{1}{2}, 1)$

28. Real-life graphs

1. (a) 30 km/h
   (b)

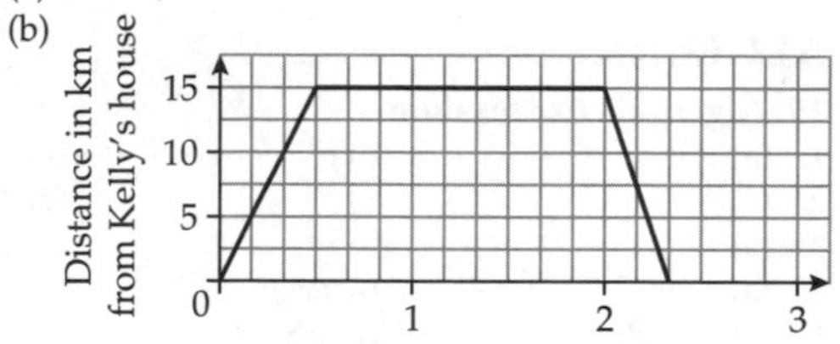

2. A and 3; B and 2; C and 4; D and 1

29. Formulae

1. 57 2. 51 3. $C = 45 + 30t$
4. $C = 6c + 4b$ 5. $P = 8x - 5$ 6. $C = (4a + 3p) \div 100$

30. Rearranging formulae

1. $t = \frac{v - u}{6}$ 2. $k = \frac{p}{2} + 3$ 3. $h = \frac{3M^2 - 1}{2}$
4. $y = \frac{5a - 6}{3 + 2a}$ 5. $p = \frac{4}{2 - t^2}$

31. Inequalities

1. (a) −2, −1, 0, 1, 2 (b) $x > 6$ 2. (a) $x > -1$ (b) $x \leqslant -9$
3. (a) $-\frac{10}{3} > y$ (b) −4 4. $-\frac{10}{3} < x < \frac{35}{3}$

32. Inequalities on graphs

1.

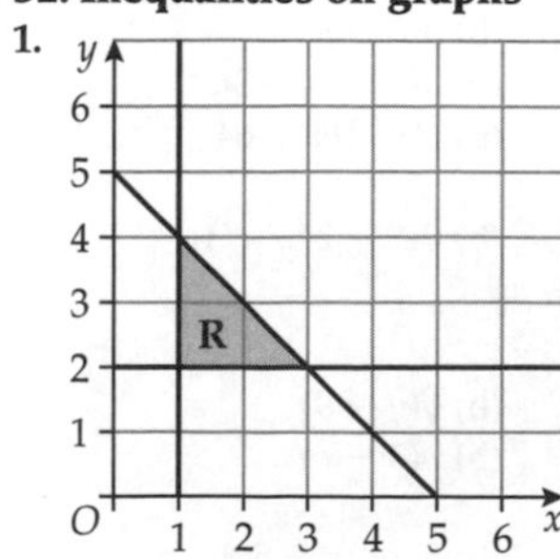

2.

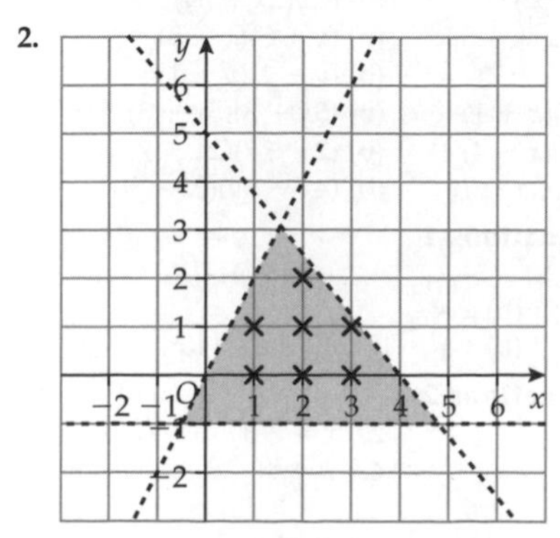

33. Quadratic and cubic graphs

1. (a)

x	−2	−1	0	1	2	3
y	−5	−6	−5	−2	3	10

(b)

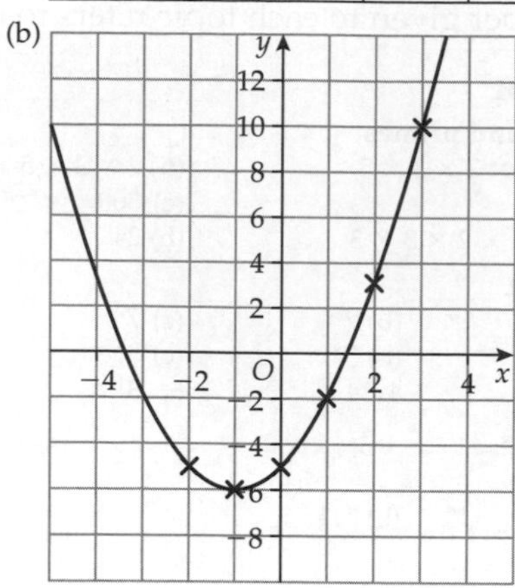

2. (a)

x	−2	−1	0	1	2	3
y	−3	1	−1	−3	1	17

(b)

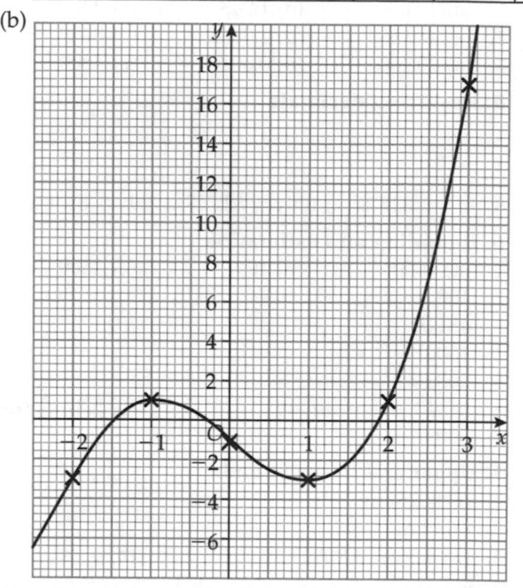

(c) −1.5, −0.35, 1.9
(d) −1.9, 0.3, 1.55

34. Graphs of $\frac{k}{x}$ and a^x

1. (a) D (b) B (c) C
2. $a = 4, k = 0.5$
3. $k = \frac{1}{8}$

35. Trial and improvement

1. 2.4
2. 3.6
3. 4.2

36. Simultaneous equations 1

1. $x = -1, y = 2$
2. $x = -0.5, y = 1$
3. $x = 1, y = 3$
4. apple 25p, banana 40p

37. Quadratic equations

1. (a) $x = 0, x = -8$ (b) $x = 9, x = -3$
2. (a) $x = 5, x = 2$ (b) $x = -1, x = -9$
3. (a) $x = -3.5, x = -3$ (b) $x = \frac{5}{3}, x = -\frac{5}{3}$
4. (a) $\frac{1}{2}(x + 1 + x + 7)(x - 2) = 72$
   $(2x + 8)(x - 2) = 144$
   $2x^2 + 4x - 16 = 144$
   $2x^2 + 4x - 160 = 0$
   $x^2 + 2x - 80 = 0$
   (b) 9 cm, 15 cm
5. $x = -\frac{5}{2}, x = \frac{1}{3}$

38. Completing the square

1. $p = 4, q = 3$
2. $a = 6, b = -41$

3. (a) $p = 1, q = 6$ (b)

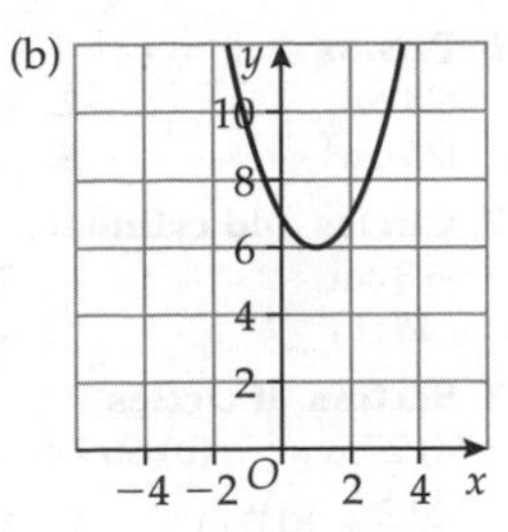

4. $a = 9, b = -3$
5. P(0, 9), Q(−4, −7)

39. The quadratic formula

1. $x = 1.85, x = -4.85$
2. $x = -1.08, x = 1.48$
3. (a) $(x + 5)(2x - 3) + (3x + 9)x = 80$
   $2x^2 - 3x + 10x - 15 + 3x^2 + 9x = 80$
   $5x^2 + 16x - 95 = 0$
   (b) $x = 3.04, x = -6.24$
   longest side = 18.1 cm

40. Quadratics and fractions

1. $x = -5, x = 3$
2. $x = -4, x = 2$
3. $x = 4.54, x = -1.54$
4. $x = -1, x = -\frac{7}{3}$

41. Equation of a circle

1.  (a)

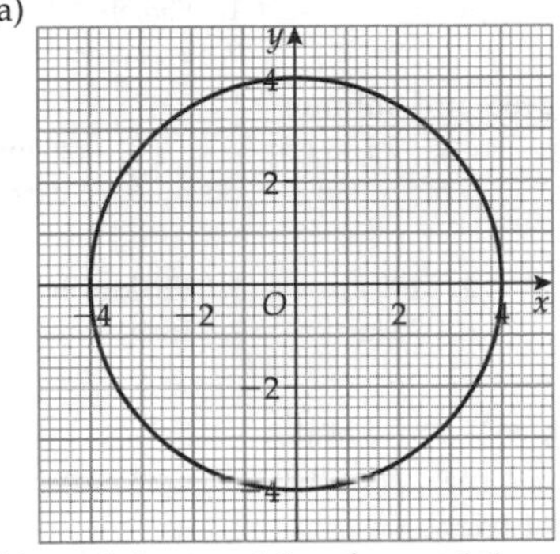

(b) $x = 3.6, y = -1.6$ and $x = -1.6, y = 3.6$

2. $x = 1.6, y = 1.3$ and $x = -2, y = -0.5$

42. Simultaneous equations 2

1. $x = -2, y = -6$ and $x = 6, y = 2$
2. $x = 6, y = 26$ and $x = -5, y = -7$
3. $x = 1, y = 4$ and $x = -2.6, y = -3.2$

43. Direct proportion

1. (a) $T = 25x$ (b) 400 N
2. 3.75 cm
3. (a) $P = 1.2s^2$ (b) 15 m/s
4. $\frac{1}{9}$

44. Proportionality formulae

1. (a) $m = \frac{8}{t^2}$ (b) $t = \frac{4}{3}$
2. $h = 480$
3. $S = 31.25$
4. $s = 3$ m/s

45. Transformations 1

1. 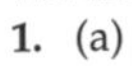 (a)

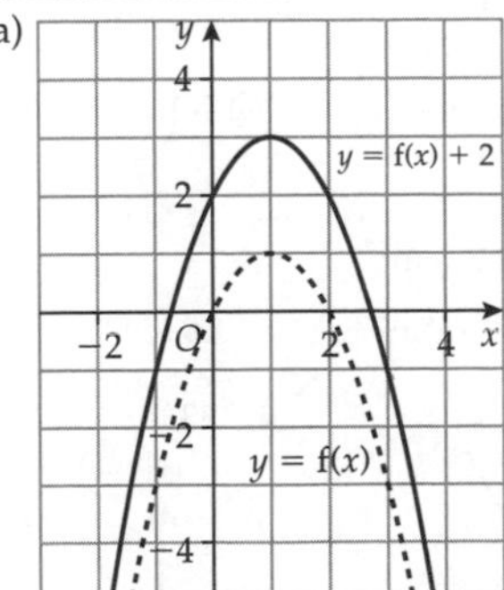

(b)

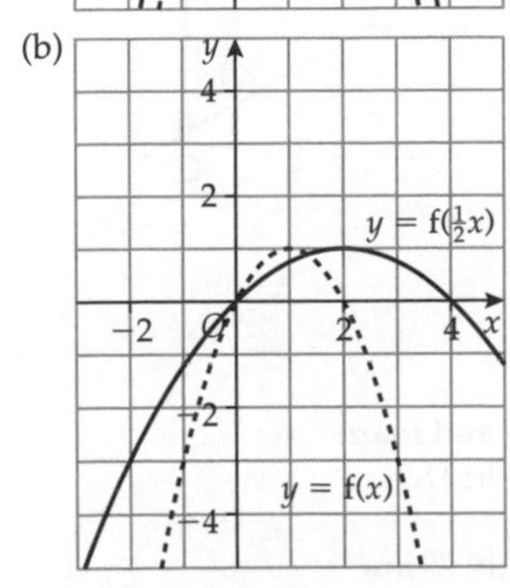

2. (a) (8, 4) (b) (−3, 4) (c) (3, 12)
3. (a)

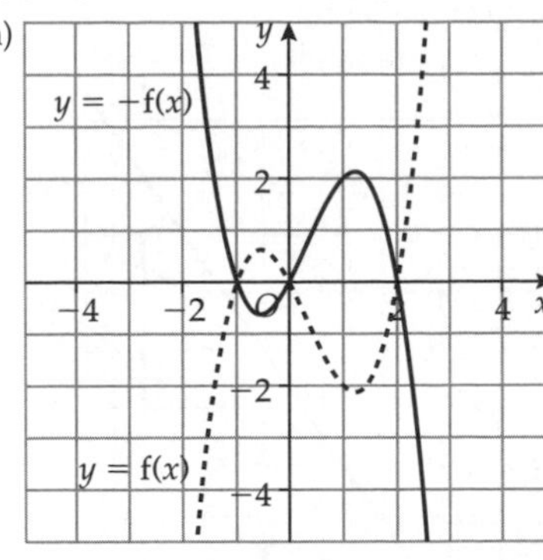

(b)

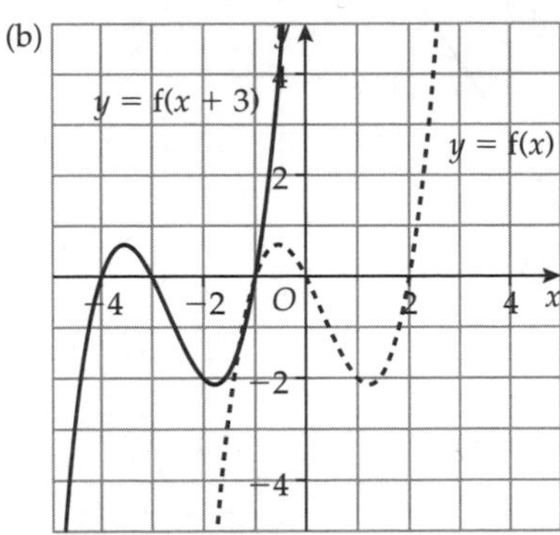

46. Transformations 2

1. (a)

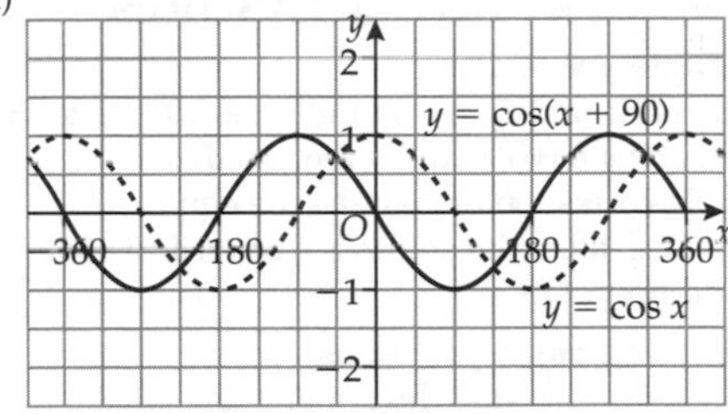

(b)

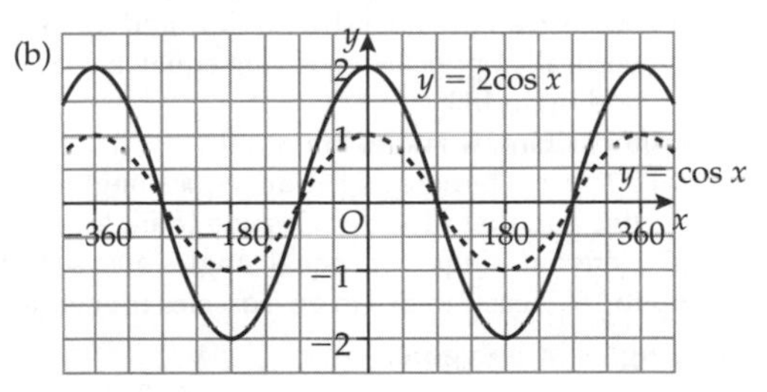

2. (a) $y = \sin(x° + 45°) - 1$
   (b) stretch scale factor 2 parallel to y-axis; translation of $\begin{pmatrix}0\\3\end{pmatrix}$

47. Algebraic fractions

1. (a) $\frac{5}{x}$ (b) $\frac{x - 5}{x - 7}$
2. (a) $\frac{1}{2x}$ (b) $\frac{x - 7}{2x + 3}$
3. (a) $\frac{11x + 7}{15}$ (b) $\frac{27x - 3}{10}$
4. $\frac{3x - 4}{(2x + 1)(x - 4)}$
5. $\frac{7x - 9}{(x + 3)(x + 5)(2x - 1)}$

48. Proof

1. e.g. $n = 3$: $5 \times 3 - 1 = 14$; 14 is not a square number
2. e.g. $2 \times 5 = 10$; 2 and 5 are prime numbers, 10 is not an odd number
3. e.g. $2n + (2n + 2) + (2n + 4) + (2n + 6) = 8n + 12 = 4(2n + 3)$
   4 is a factor so sum is always a multiple of 4 for all integer values of n
4. e.g. $(2n - 1) + (2n + 1) + (2n + 3) = 6n + 3 = 3(2n + 1)$
   3 is a factor so sum is a multiple of 3 for all integer values of n
5. $16n^2 + 8n + 1 - 16n^2 + 8n - 1 = 16n = 8(2n)$
   8 is a factor so expression is a multiple of 8 for all positive integer values of n

49–50. Problem-solving practice

1. 89°
2.

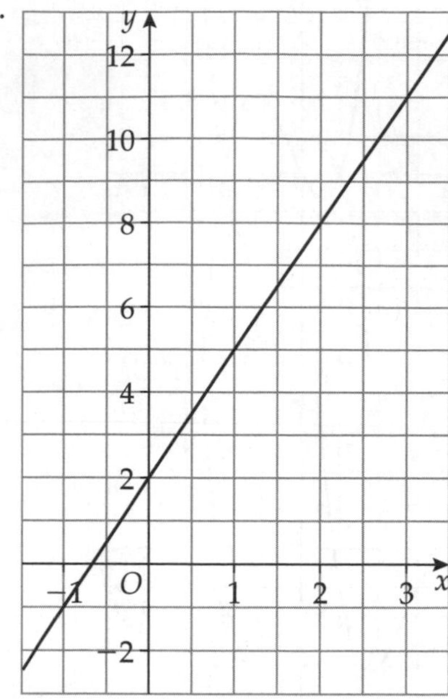

3. B has coordinates (6, −7, 2)
4. Proof
5. (a) Translation of $\begin{pmatrix}3\\0\end{pmatrix}$
   (b) Reflection in the x-axis

GEOMETRY AND MEASURES

51. Angle properties

1. (a) 142°; angles on a straight line add up to 180°
   (b) 38°; alternate angles are equal
2. 126°; corresponding angles are equal
3. 66°; e.g. alternate angles are equal and angles in a triangle add up to 180°

52. Solving angle problems

1. (a) 135°; corresponding angles are equal
   (b) 50°; e.g. angles on a straight line add up to 180°; vertically opposite angles are equal; angles in a triangle add up to 180°
2. (a) 85°; alternate angles are equal
   (b) 30°; e.g. alternate angles are equal; angles on a straight line add up to 180°; angles in a triangle add up to 180°
3. 37°; alternate angles are equal; angles on a straight line add up to 180°; base angles in an isosceles triangle are equal

53. Angles in polygons

1. 36°   2. 24°   3. 8
4. 18   5. 165°
6. 36°; e.g. exterior angles of a polygon add up to 360°; angles on a straight line add up to 180°; base angles of an isosceles triangle are equal; angles in a triangle add up to 180°

54. Plan and elevation

1.

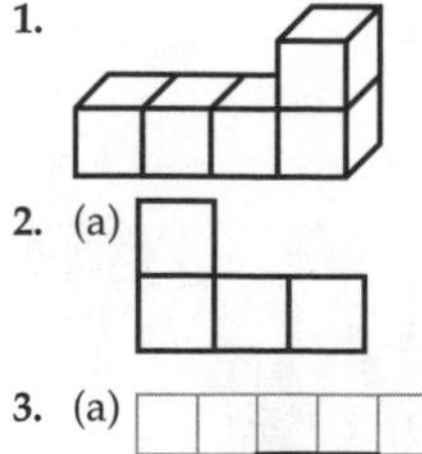

2. (a) (b)

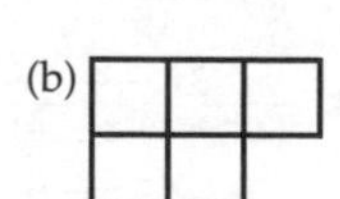

3. (a) (b)

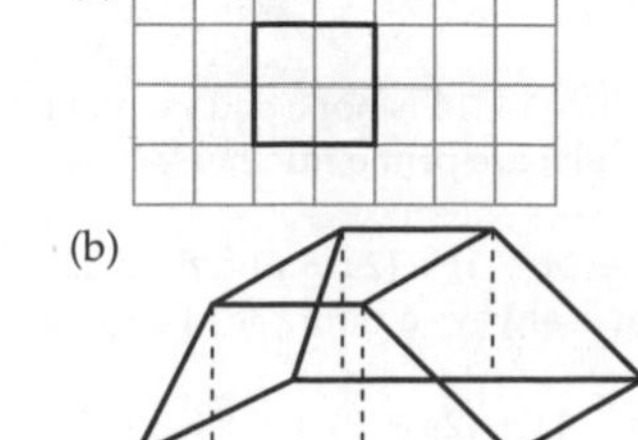

55. Perimeter and area

1. 72 cm²   2. (a) 42 cm² (b) 76 cm²   3. 6 tins
4. Yes – he wanted at least £41 900

56. Prisms

1. 228 cm²   2. 408 m²   3. 360 cm³
4. 320 cm³   5. 1520 m³

57. Circles and cylinders

1. 46.3 cm   2. (a) 1850 cm³ (b) 836 cm²
3. 4.15 cm   4. 25π cm²

58. Sectors of circles

1. (a) 275 cm² (b) 66.7 cm
2. $\left(\frac{35\pi}{18} + 10\right)$ cm   3. 28.0 cm²

59. Volumes of 3-D shapes

1. 288π cm³   2. 1820 cm³
3. 86.9 cm³   4. 264π cm³
5. 385 cm³

60. Pythagoras' theorem

1. 10.5 cm   2. 24 cm   3. 20.2 cm   4. 22.2 cm   5. 11.1 cm

61. Surface area

1. 1260 cm²   2. 415 cm²
3. $\frac{32}{3}$ cm   4. 9 cm

62. Converting units

1. 3 hours and 40 minutes   2. £233.10
3. Shop B   4. £58.50
5. 2 hours   6. 5 glasses

63. Units of area and volume

1. 7000 mm³   2. 75 000 cm²   3. 420 m²
4. 6000 litres   5. 50 cm   6. 6 cartons

64. Speed

1. 300 km/h   2. 200 km   3. 120 km/h   4. 1350 km
5. 4 km   6. 30 km   7. 16:40

65. Density

1. (a) 7130 kg/m³ (b) 13 440 kg
2. 129.6 g   3. 4.632 kg   4. 0.65 g/cm³

66. Congruent triangles

1. angle CAB = angle ACD (alternate angles)
   angle BCA = angle DAC (alternate angles)
   AC is common
   So triangle ABC is congruent to triangle CDA by AAS
2. angle AED = angle CDE (both interior angles of regular pentagon)
   $AE = CD$ (side of regular pentagon)
   ED is common
   Congruent by SAS
3. AD is common
   $AB = AC$ (sides of an isosceles triangle are equal)
   angle ADC = angle ADB = 90° (given)
   Congruent by RHS

67. Similar shapes 1

1. (a) 9 cm (b) 6.67 cm
2. 8 cm   3. 9.8 cm

68. Similar shapes 2

1. 4160 cm³
2. (a) 10 cm (b) 128 cm³
3. 171 cm²

69. Bearings

1. 330°   2. 320°
3.

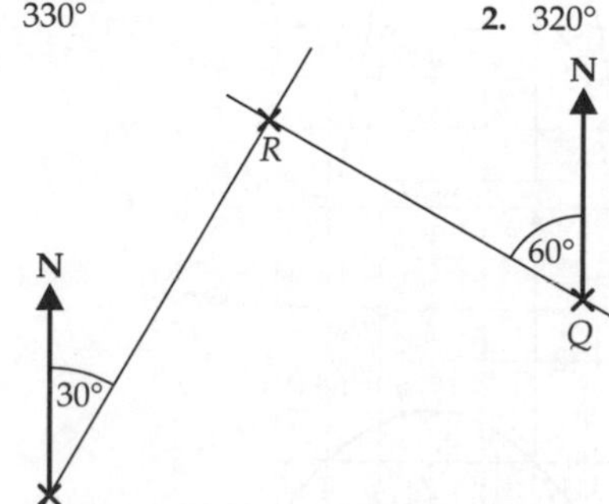

70. Scale drawings and maps

1. (a) 71 m (b) 11.9 cm
2. 64 km
3. (a) 35 km (b) 29 km

71. Constructions

1.

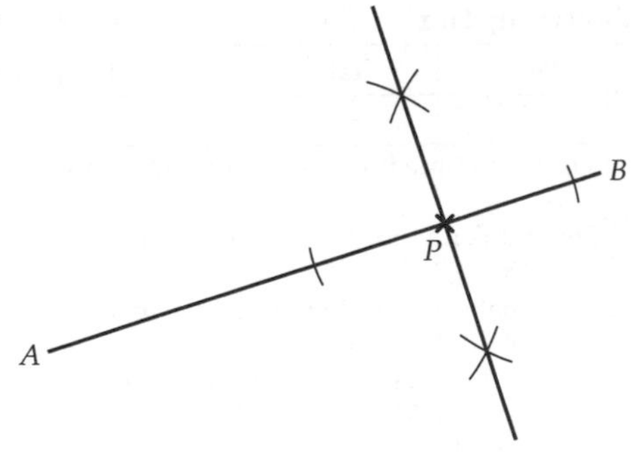

2.

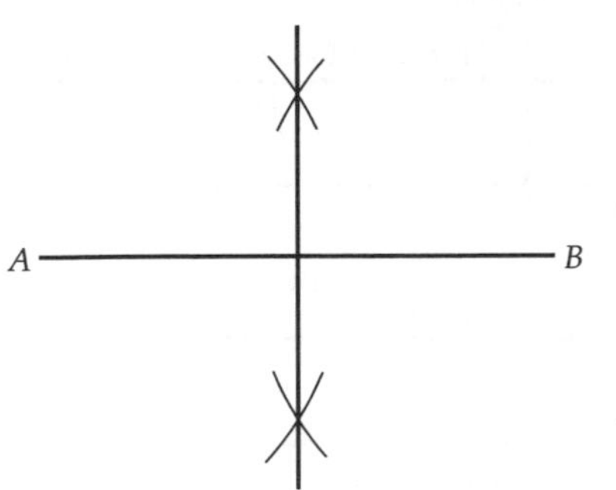

3.

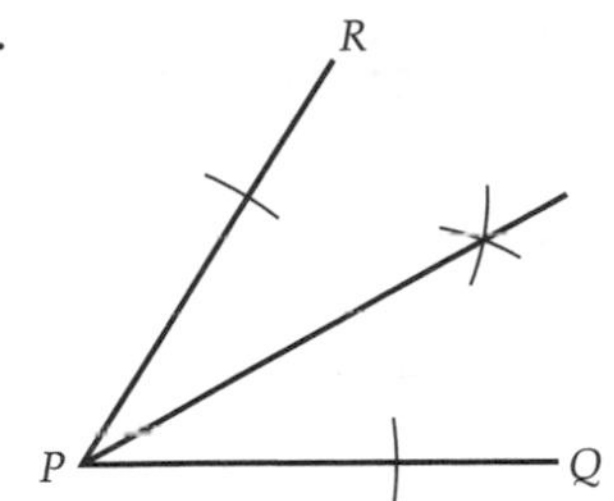

72. Loci

1.

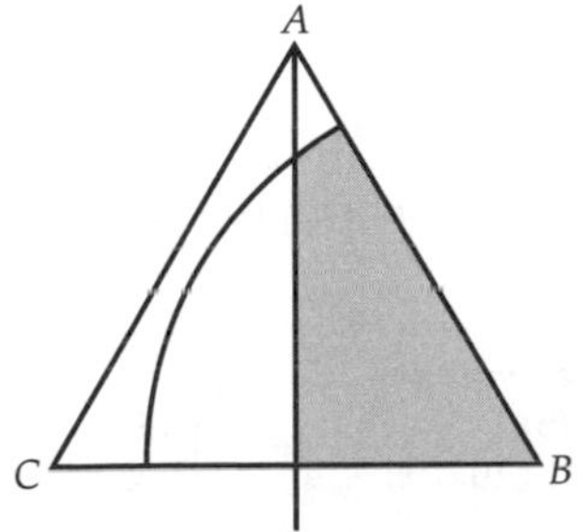

2.

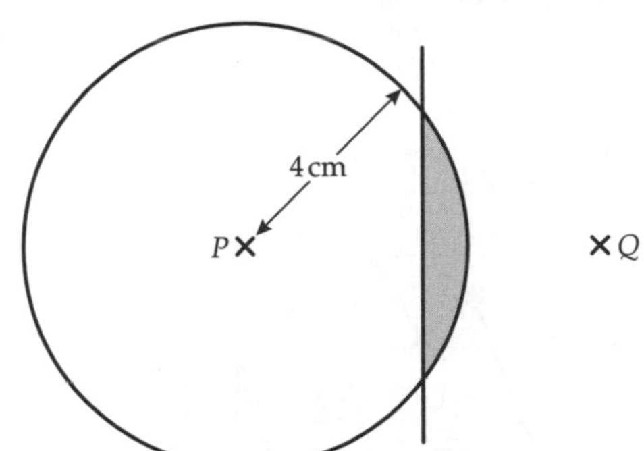

3. A B D C 2 cm 3.5 cm

73. Translations, reflections and rotations

1.

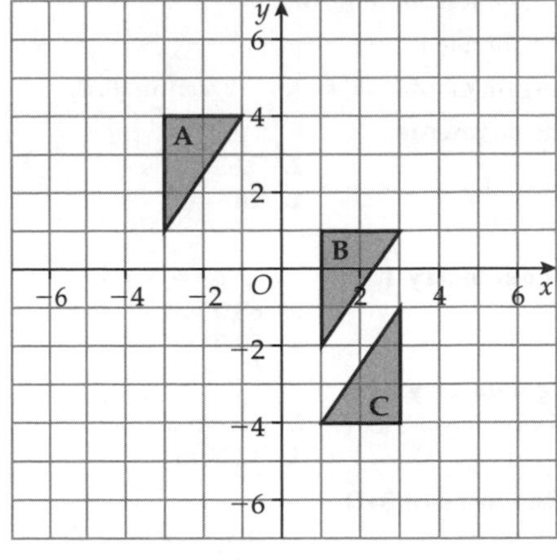

2. (a) translation $\begin{pmatrix} -5 \\ 2 \end{pmatrix}$

(b), (c)

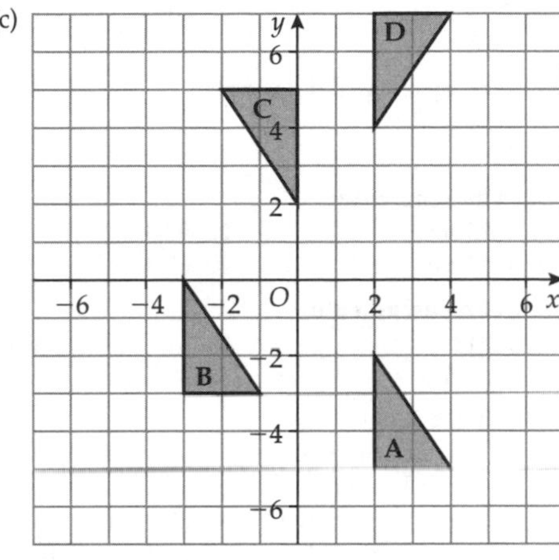

3. (a) reflection in line $y = x$

(b) 90° clockwise rotation about origin

74. Enlargements

1.

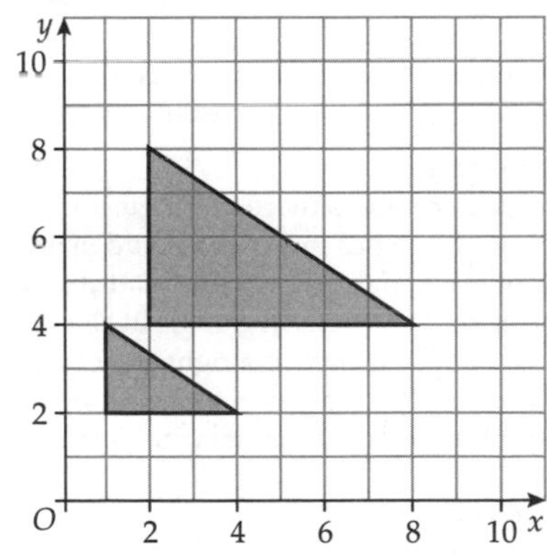

2. enlargement of scale factor 3, centre (1, 0)

3. (a), (b)

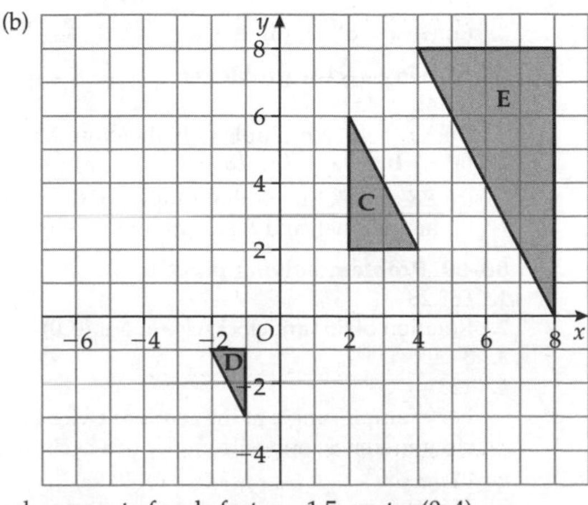

(c) enlargement of scale factor -1.5, centre (0, 4)

75. Combining transformations

1. 180° rotation about (4, 0)
2. translation $\begin{pmatrix}0\\6\end{pmatrix}$
3. enlargement of scale factor −2, centre (0, 0)

76. Line segments

1. 13
2. 5.83
3. 15
4. 26
5. 11.2

77. Trigonometry 1

1. 58.9°
2. 68.0°
3. 55.9°
4. 59.0°

78. Trigonometry 2

1. 9.34 cm
2. 6.1 cm
3. 10.0 cm
4. 12.8 cm

79. Pythagoras in 3-D

1. 18.0 cm
2. diagonal = 27.7 cm
3. 15.2 cm
4. 9.58 cm

80. Trigonometry in 3-D

1. 69.9°
2. 31.6°
3. 26.5°

81. Triangles and segments

1. 22.4 cm²
2. 27.9 cm²
3. 39.0°
4. 85.6 cm²

82. The sine rule

1. 8.10 cm
2. 92.3°
3. 39.0 cm
4. 65.3 cm²

83. The cosine rule

1. 7.54 cm
2. (a) 34.8° (b) 47.9 m²
3. 41.6 cm
4. 4210 m²

84. Circle facts

1. 17.5°; e.g. angle between tangent and radius is 90°; angles in a quadrilateral add up to 360°; tangents from a point outside a circle are the same length
2. 46°; e.g. angle between tangent and radius is 90°; tangents from a point outside a circle are the same length; angles in a triangle add up to 180°; base angles of an isosceles triangle are equal
3. 63°; e.g. angle between tangent and radius is 90°; angles in a triangle add up to 180°; base angles of an isosceles triangle are equal
4. 29°; angle between a radius and a tangent is 90°; isosceles triangle; base angles of an isosceles triangle are equal

85. Circle theorems

1. 15°; e.g. angle at centre is twice the angle at circumference, isosceles triangle; angles in a triangle add up to 180°
2. 132°; e.g. opposite angles of a cyclic quadrilateral add up to 180°, angle at centre is twice the angle at circumference
3. 40°; e.g. angle in a semicircle is a right angle; angles in a triangle add up to 180°; angles in the same segment are equal
4. 62°; e.g. alternate segment theorem; angles in a triangle add up to 180°

86. Vectors

1. (a) $\begin{pmatrix}3\\4\end{pmatrix}$ (b) (10, 9)
2. (a) $\mathbf{b} - \mathbf{a}$ (b) $\frac{3}{5}\mathbf{b} + \frac{2}{5}\mathbf{a}$
3. (a) (i) $\mathbf{a} + \mathbf{d}$ (ii) $\mathbf{d} - \mathbf{a}$ (b) $\frac{\mathbf{a} + \mathbf{d}}{2}$

87. Solving vector problems

1. (a) $4\mathbf{b} - 4\mathbf{a}$
   (b) $\overrightarrow{XY} = 4\mathbf{b}$, a multiple of **b**, therefore *XY* is parallel to *OR*
2. (a) (i) $\mathbf{b} - 3\mathbf{a}$ (ii) $2\mathbf{b} - 2\mathbf{a}$ (iii) $\frac{1}{2}\mathbf{b} - \frac{1}{2}\mathbf{a}$
   (b) $\overrightarrow{PR}$ and $\overrightarrow{PQ}$ are both multiples of $(\mathbf{b} - \mathbf{a})$ so *PR* and *PQ* are parallel, and *P* is a common point.

88–89. Problem-solving practice

1. £68.28
2. Rotation of 90° anticlockwise about (0, 0)
3. 37.1°
4. 128°
   For example, angle at the centre is twice the angle at the circumference; opposite angles of a cyclic quadrilateral add up to 180°.
5. 116 m

STATISTICS AND PROBABILITY

90. Collecting data

1.

Pet	Tally	Frequency

2. Two from: no time frame, overlapping response boxes, no zero
3. (a) Two from: no time frame, options not specific enough, no zero
   (b) How many emails do you receive in a day?
   0–10 ☑ 11–20 ☑ 21–30 ☑ more than 30 ☑
4. How often do you go on holiday each year?
   0 ☑ 1 ☑ 2 ☑ 3+ ☑

91. Two-way tables

1.

	France	Germany	Spain	Total
Female	12	13	13	38
Male	21	12	9	42
Total	33	25	22	80

2.

	School lunch	Packed lunch	Other	Total
Female	16	6	17	39
Male	9	8	4	21
Total	25	14	21	60

3.

	Boys	Girls	Total
Liked coffee	13	11	24
Did not like coffee	2	14	16
Total	15	25	40

4.

	Jazz	Country	Other	Total
Male	30	8	9	47
Female	2	2	24	28
Total	32	10	33	75

92. Stratified sampling

1. 14
2. 12
3. 34
4. 18

93. Mean, median and mode

1. (a) 4 (b) 5.5
   (c) 9.6 (d) Student's explanation
2. 14.5
3. £301

94. Frequency table averages

1. (a) 0 (b) 1 (c) 1.15
2. (a) $10 < m \leqslant 20$ (b) $10 < m \leqslant 20$ (c) 15.25 minutes

95. Interquartile range

1. (a) 24 (b) 13
2. (a) 31 (b) 18
3. 25

96. Frequency polygons

1.

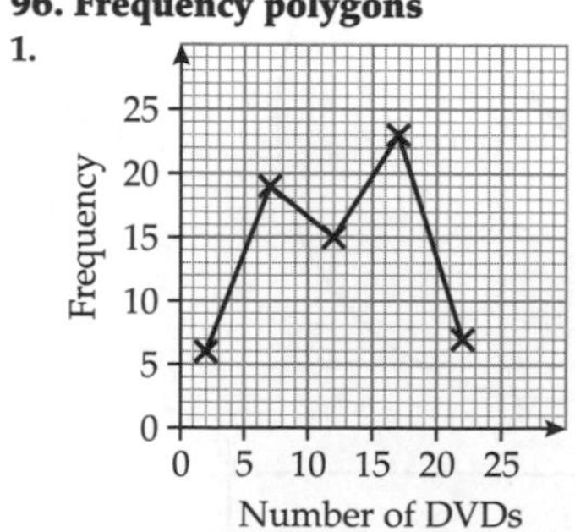

2.

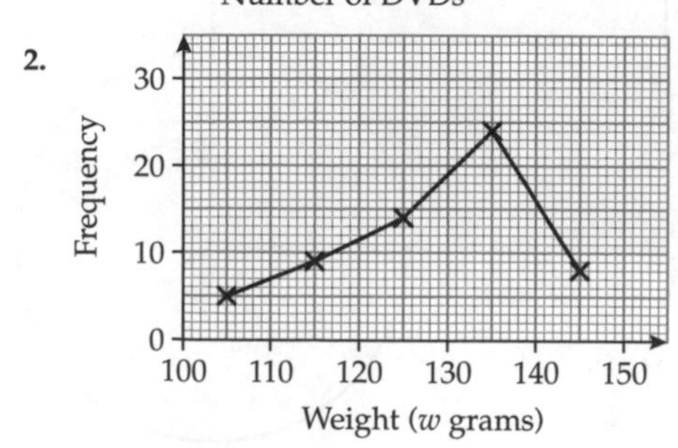

97. Histograms

1. (a)

Time (minutes)	Frequency
$5 < x \leqslant 10$	15
$10 < x \leqslant 15$	20
$15 < x \leqslant 25$	60
$25 < x \leqslant 35$	50
$35 < x \leqslant 50$	15

(b)

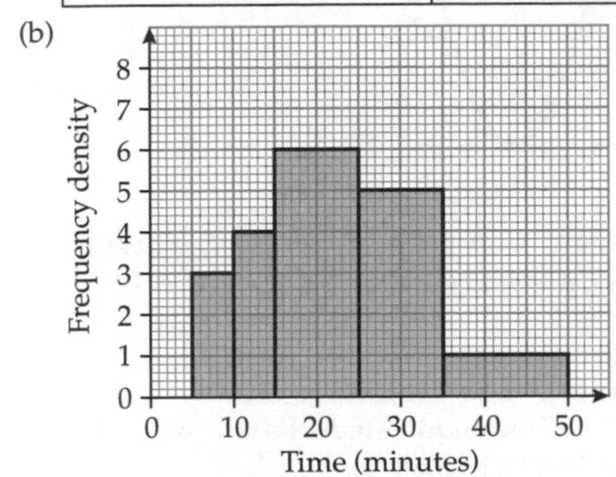

2. (a)

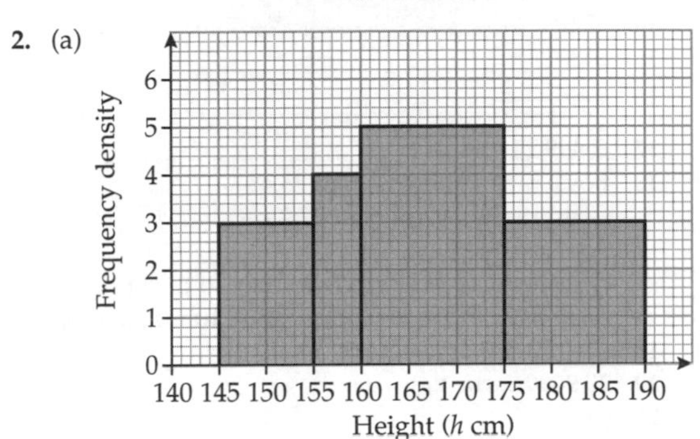

(b) 70

98. Cumulative frequency

1. (a) 25 (b) 12
2. (a)

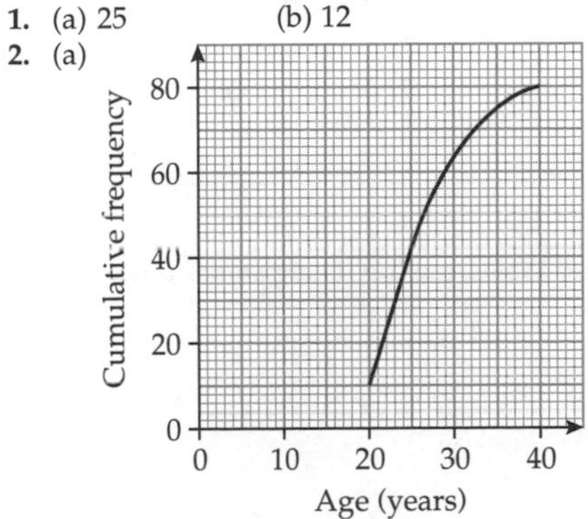

(b) 7

(c) No: $\frac{28}{80}$ people were older than 27, which is 35%.

99. Box plots

1. (a) 162 cm (b) 31 cm (c) 30
2. Jean is incorrect. 25% of the cases fall between the median and upper quartile.
3.

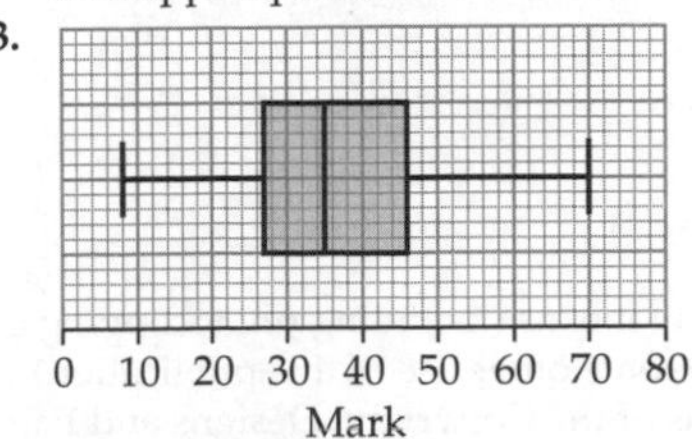

100. Scatter graphs

1. (a) positive (b) line of best fit drawn (c) 20 or 21
2. (a) negative (b) about £4000 (c) about 3 years

101. Probability

1. 0.2
2. 18
3. (a) 0.35 (b) 30
4. 108

102. Tree diagrams

1. (a) $\frac{3}{4}, \frac{3}{5}, \frac{2}{5}, \frac{3}{5}$ (b) $\frac{11}{20}$
2. $\frac{1}{3}$
3. $\frac{13}{18}$

103–104. Problem-solving practice

1. 15
2. Tree A: range 27 g, median 53 g
   Tree B: range 22 g, median 56.5 g
   For example:
   Tree B has heavier apples as the median is higher.
   Tree A has larger range which means more variety in the weights.
3. 61
4. Katie is wrong because 25% of all the bags have a weight greater than 19.8 kg.
5. 130
6. $\frac{13}{28}$

106–112. Paper 1 Practice exam paper

1. 110°
   Base angles in an isosceles triangle are equal; angles in a triangle add up to 180°; angles on a straight line add up to 180°.
2. (a) 0.16 (b) 32
3. Enlarged shape
4. (a) $12x - 23y$
   (b) $4y(3y + 2)$
   (c) $x^2 - 2x - 48$
5. (a) $6n + 1$
   (b) For example:
   No, 104 is an even number and all the terms in the sequence are odd numbers.
6.

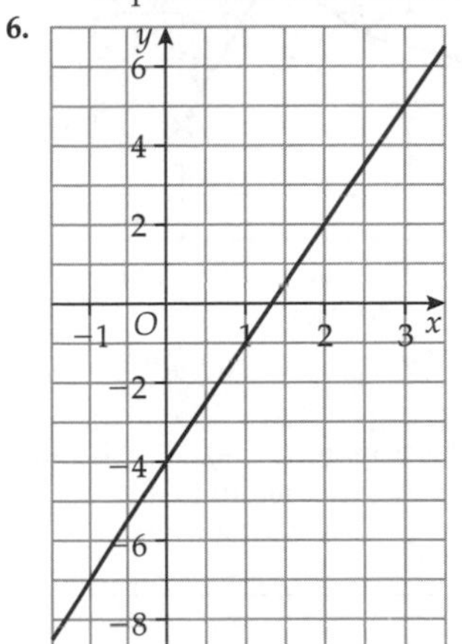

7. £169
8. (a) For example:
   How many magazines do you buy each week?
   ☐ 0 ☐ 1 ☐ 2 ☐ 3 ☐ 4 or more
   (b) For example:
   He only gives the questionnaire to women, not to men.
   He only gives the questionnaire to people coming out of the newspaper shop.
9. (a) 6 packs of bread rolls, 9 packs of burgers (b) 216
10. (a) $y = \frac{18}{5}$ (b) $x = \frac{17}{2}$ (c) $(3x + 1)(2x - 5)$
11. (a) 17.5 m (b) 18.5 m
12. 30 000 cm²
13. $x = 3, y = -2$
14. Arc radius 7 cm, centre B: perpendicular bisector of AB: shaded region enclosed by arc and line
15. $1\frac{1}{5}$
16. £43
17. (a) 1 (b) ±8 (c) $\frac{1}{16}$
18. (a) £43 (b) £30 (c) 8
    (d) The interquartile range is not affected by any outliers.
19. 118°
    For example, angle at the centre is twice the angle at the circumference; opposite angles of a cyclic quadrilateral add up to 180°.
20. $y = -\frac{8}{3}x + 3$
21. Proof
22. $\frac{224}{495}$
23. (a) $3\mathbf{a} + \mathbf{c}$ (b) $\mathbf{a} + \mathbf{c}$
24. (a) All have an equal chance of being chosen.
    (b) The number from each group is proportional to the number in the population.

25. (a)

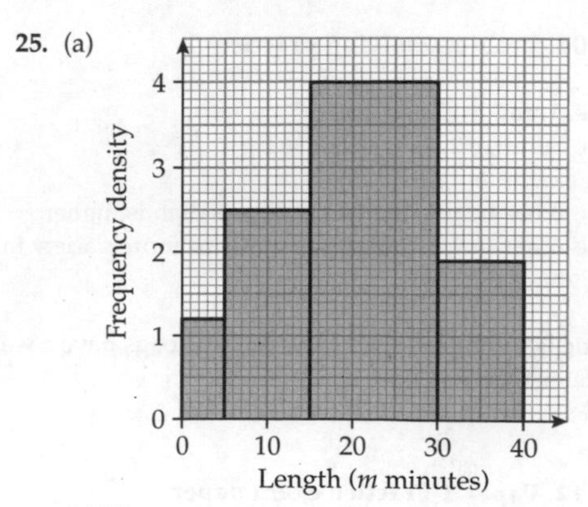

(b) 59

26. $h = \frac{32}{9}x$

113–118. Paper 2 Practice exam paper

1. 300 g flour, 5 eggs, 500 ml milk, 125 g butter
2. 85p
3. (a) 10.872 689 85
   (b) 10.9
4.

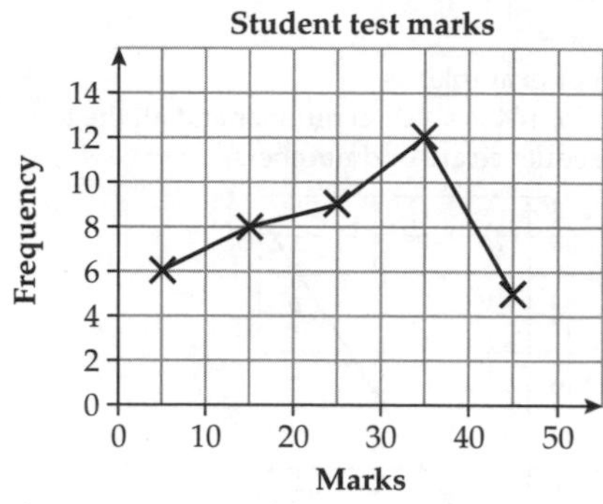

5. (a) x^{11} (b) y^6 (c) m^{14}
6. £38.88
7. (a) Triangle with vertices at (1, −4), (3, −4) and (1, −7)
   (b) Rotation of 90° clockwise about (0, 0)
8.

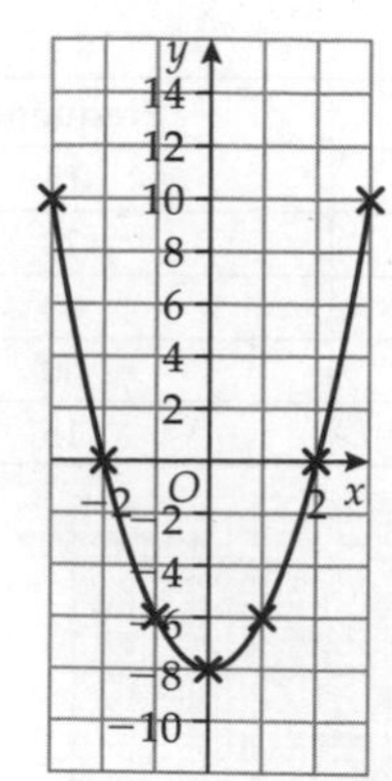

9. (a) 9.02×10^{-3} (b) 608 000 (c) 117 000
10. Yes, it is long enough; it will reach to 5.83 m.
11. 11 tins
12. 5.6
13. (a) $40 < t \leq 50$ (b) 50.5 seconds
    (c) No, the middle student's time will still be in the $40 < t \leq 50$ interval.
14. 32.5°
15. £480
16. $a = \frac{t + 8}{4 + 2t}$
17. £6499.24
18. Boys: range 4.6 s, median 28.4 s
    Girls: range 4 s, median 27.7 s
    For example:
    Girls have a lower median time, so they finished quicker.
    Boys have a larger range so their times are more varied.
19. $x = -0.180$ or $x = 1.85$
20. 20.9090… cm
21. 0.6
22. 737 m²
23. 41
24. $\frac{x - 3}{2x + 1}$
25. $\frac{8}{11}$
26. (a) (−2, 3) (b) (1, 5) (c) (−2, 5)

Published by Pearson Education Limited, a company incorporated in England and Wales, having its registered office at Edinburgh Gate, Harlow, Essex, CM20 2JE. Registered company number: 872828

www.pearsonschoolsandfecolleges.co.uk

Edited by Fiona McDonald and Laurice Suess
Typeset by Tech-Set Ltd, Gateshead

First published 2011

17 16 15 14
15 14 13 12

British Library Cataloguing in Publication Data
A catalogue record for this book is available from the British Library

ISBN 978 1 44690 015 4

Printed in Slovakia by Neografia